SCREEN ADAPTATION

SCREEN ADAPTATION

A Scriptwriting Handbook

Second Edition

Dr. Kenneth Portnoy

Focal Press

Boston Oxford Johannesburg Melbourne New Delhi Singapore

Focal Press is an imprint of Butterworth–Heinemann.

A member of the Reed Elsevier group

Recognizing the importance of preserving what has been written, Butterworth–Heinemann prints its books on acid-free paper whenever possible.

Butterworth–Heinemann supports the efforts of American Forests and the Global ReLeaf program in its campaign for the betterment of trees, forests, and our environment.

Library of Congress Cataloging-in-Publication Data

Portnoy, Kenneth, 1946–
Screen adaptation : a scriptwriting handbook / Kenneth Portnoy. — 2nd ed.
p. cm.
Includes bibliographical references and index.
ISBN 0-240-80349-3 (alk. paper)
1. Motion picture authorship. I. Title.
PN1996.P67 1998
808.2'3—dc21 98-18039
CIP

British Library Cataloguing-in-Publication Data
A catalogue record for this book is available from the British Library.

The publisher offers special discounts on bulk orders of this book.
For information, please contact:

Manager of Special Sales
Butterworth–Heinemann
225 Wildwood Avenue
Woburn, MA 01801-2041
Tel: 781-904-2500
Fax: 781-904-2620

For information on all Butterworth–Heinemann publications available, contact our World Wide Web home page at: http://www.bh.com

10 9 8 7 6 5 4 3 2 1

Printed in the United States of America

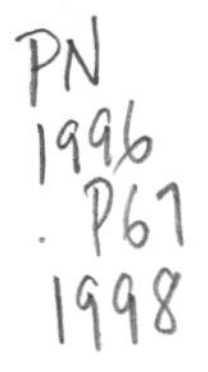

Contents

Preface

Since the publication of the first edition of this book, several successful screenplays, the origins of which lie in a novel, short story, or play, have been produced. Among these screenplays are *The Bridges of Madison County*, *Jurassic Park*, *Primal Fear*, *Driving Miss Daisy*, *Dances with Wolves*, *The Silence of the Lambs*, *The Shawshank Redemption*, *Hunt for Red October*, *The Firm*, *Apollo 13*, *Leaving Las Vegas*, *The English Patient*, *Marvin's Room*, *Hamlet*, *Contact*, *Portrait of a Lady*, and *Romeo and Juliet*. These are only a few of the many stories that have successfully made the transition from narrative fiction to the screen.

This edition examines these new works along with previous examples discussed in the first edition. Through this exploration process and various new writing exercises, I hope to focus on the problems inherent in the adaptation process. With half of all new screenplays based on material adapted from another writer's work, this process is of vital interest to any writer who hopes to write his or her own screenplays.

The material in this book is derived from my many years as a professor in the Radio, TV, and Film's writing program at California State University, Northridge, located in a suburb of Los Angeles. The Radio–TV–Film Department has a rather large program with several sections of introductory media writing required of all majors. The introductory course examines media writing and the various formats for movie, sitcom, episodic, and corporate scriptwriting. The intermediate course is devoted to expanding the lessons of the introductory course, culminating in the writing of a 30-minute script based on an original idea. Most students base their story on a newspaper or magazine article, or even personal experience.

Our advanced writing course requires the completion of a 60-minute script based on material adapted from a novel or short story. Our final course requires students to write a 120-page full-length script based on original or adapted material. In addition, Northridge offers courses in comedy writing, corporate scriptwriting, and a master's program in scriptwriting.

The rationale for the course sequence is simple. The writing instructors at Northridge have always felt that the development of an original script is a prerequisite for the more advanced problems of adapting another writer's work. If students are able to do a successful adaptation, they must call on the skills of the original material's writer. Students must be prepared to alter and develop new story lines, add characterization, develop conflict with characters, and deal with a relevant social issue using realistic dialogue and well-constructed scenes. Most students fail at this task! Many become so enamored of their novels or short stories that they forget the lessons about development of character and careful story construction they learned in the beginning writing course. Because of their obsession to remain faithful to the original writer, they forget that they are creating a new work, developing a new genre, and creating a new art form.

This book has grown out of many years of looking for new ways to teach the advanced writing adaptation course. It looks at the specific dramatic skills one needs to develop a successful screenplay. It also explores how and why other screenwriters have changed the adapted works to make their own screenplay work. Therefore, this book is more than a text in screen adaptation; it is designed for all people interested in the writing process. Unlike other screenwriting books, it looks at the principles of dramatic writing from a slightly different perspective. By examining the choices and processes of other screenwriters, this work tries to unlock some of the mystery involved in the development of a successful screenplay.

Whether one develops an original story idea or adapts the work of another writer, the same dramatic principles are necessary for the successful conceptualization of the screenplay. Through the examples, discussions of the problems of adaptation, and the many exercises given in this book, it is my hope that the beginning or advanced writer will gain fresh insight into the screenwriting process.

Above all, this book explores the concept of drama. But what is drama? Perhaps Paddy Chayefsky (1954, p. 178) said it best in the introduction to one of his early teleplays, *Marty*:

> I do not like to theorize about drama. I suspect the academic writer, the fellow who can precisely articulate his theater. However, it is my current belief that the function of the writer is to give the audience some shred of meaning to the otherwise meaningless patterns of their lives. Our lives are filled with endless moments of

> stimulus and depression. We relate to each other in an incredibly complicated manner. Every fiber of relationship is worth a dramatic study. There is far more exciting drama in the reasons why a man gets married than in why he murders someone. The man who is unhappy in his job, the wife who thinks of a lover, the girl who wants to get into television, your father, mother, sister, brothers, cousins, friends—all these are better subjects for drama than Iago. What makes a man ambitious? Why does one girl always try to steal her kid sister's boy friends? . . . These are the substances of good television drama; and the deeper you probe into and examine the twisted, semiformed complexes of emotional entanglements, the more exciting your writing becomes.

The purpose of this book is to help you find fresh ways to make your writing exciting to yourself and others. If you base your screenplay on an original idea, then maybe my examples will remind you of the dramatic elements necessary for an exciting, well-written screenplay. If you choose to adapt another person's work, then maybe this book will help you avoid some of the simple pitfalls into which all adapters are likely to fall.

1 Introduction

Many interesting films have been developed from short stories, novels, and plays. But the process of adapting a work into a screenplay is complex and often fraught with tricky problems. The entire novel may require several readings. The screenplay, however, must tell a story within a two-hour framework or risk losing the audience's attention and the profitability factor so crucial to a successful film, TV, or cable dramatization.

Dune, for example, is a 500-page novel. But the modern film audience doesn't want to sit through an epic film. As a result, the writer condensed the story into two hours and ten minutes, losing many of the key elements that made the novel so popular with its readers. *The Big Sleep* is another film from a novel, but the movie underwent several crucial changes to accommodate audiences of the late 1940s. Elements dealing with homosexuality, drugs, nudity, and pornography were eliminated from the film to avoid offending the more conservative audience of the time.

When stories are too long, it becomes necessary to reduce the number of subplots and characters to fit the two-hour storytelling length. *East of Eden*, a John Steinbeck novel, was adapted for the screen. Although it had several plot and subplot lines, the adapter concentrated on only one plot line—the story of the Trask family. Why was this plot line chosen? Because the Trask family's story had the most conflict—a key element of any successful screenplay.

STRUCTURE

Adaptations always change structure. In *The Bridges of Madison County*, the focus of the novel is entirely changed by making Francesca the central character instead of Robert. Now the story has a more typical dramatic structure with a protagonist, a strong inner conflict, and the possibility of character growth.

In *The English Patient*, the screenwriter also changed the focus of his main characters. In the novel, the story focuses on Hana's relationship with the Indian soldier and her uncle, the spy. In the movie, an English patient's love affair becomes the main plot line, with Hana and her relationship with the soldier forming the subplot.

In *Leaving Las Vegas,* the O'Brien novel (1990) begins with Sera and the backstory about how she became a prostitute. In the movie, Ben and his drinking problem become the central focus illustrated visually in the opening scene at a supermarket. Ben, wheeling a shopping cart, stops to buy a bottle of vodka, then scotch, then brandy, then beer, and so on.

In *The Grapes of Wrath*, another Steinbeck novel adapted for the screen, key time-sequence elements were altered to accommodate a more dramatic structure that would build to a climax and create more tension and suspense. In the novel, for example, the Joads experience the friendly government camp at the beginning of the book after finally arriving in California. In the screenplay, this incident occurs later in the story, thereby allowing the Joads' problems to intensify as their search for work and their negative treatment in California portrays even more dramatically the hopeless plight of these migrant people.

In *The Ox-Bow Incident*, the entire novel's ending was changed to create a more visual filmic experience. Because of the screenplay's limited time sequence and the fact that the viewers cannot go back and review material from the original work, the screenwriter must have a progressively advancing story line with explicit conflicts that allows the viewer some understanding of the writer's message.

Sometimes, as in *The French Lieutenant's Woman*, a subplot is added to achieve audience identification and understanding. Here, the movie company subplot was added so that the audience has a more modern element to identify with in terms of character conflict and problems. By updating these problems to a twentieth-century movie company, rather than the nineteenth-century landscape of the novel, the screenwriter succeeds in giving the audience a more recognizable terrain. Thus, the movie becomes more of a relationship story, which is so popular with twentieth-century audiences.

A novel's structure is sometimes condensed to speed the story up. Such is the case in the film version of *The Graduate* (Henry, 1967). In the novel, the climax involves Ben telephoning Elaine, pacing in front of her dorm, looking for her in the cafeteria, and searching for her in the library before his father suddenly appears. At this point, an entire scene is devoted to Ben and his father. Ben explains the sordid details between himself and Mrs. Robinson. His father wants him to see a shrink. Ben refuses and then goes off to San Francisco to find Carl and Elaine. Ben finally winds up in Santa Barbara for the wedding; the novel ends with Ben taking Elaine away before she marries Carl.

In the screenplay, these scenes are condensed. The climax focuses on Ben getting to Santa Barbara to stop the wedding; the screenwriter puts in some amusing obstacles, such as Ben running out of gas, to prolong the tension. When Ben finally arrives at the church, Elaine has just taken her wedding vows. This provokes Ben into his chant of pain and causes Elaine to leave Carl standing at the altar. The entire scene is more dramatic than it is in the novel. It builds to a nice pitch with Ben arriving as the vows are completed. The drama is reinforced by the scenes of Ben running to get to the church before the end of the ceremony.

CHARACTERIZATION

Characterization is a key element in both novels and screenplays. But, in the screenplay, the writer has fewer pages to establish and make the characters sympathetic. In *Heartburn*, for example, the crucial argument between Rachael and Mark at their friends' house is added into the screenplay. Their argument over a piece of chicken reinforces that Rachael is far from the perfect woman, unlike the Rachael in the novel. She gets angry, fights in public, and leaves the party in a fit. Movie audiences aren't interested in perfect characters. They want to see complex human beings like themselves. The scene is also more dramatic—it has conflict and intensity. It holds the audience's attention.

Sometimes a character in a source story is too negative. In *The Big Sleep*, the character of Vivian is too sleazy for the protagonist, Marlowe, to fall in love with. Therefore, the writer changes the character and makes her more well-rounded. After a shoot-out with a gangster during the film's climax, Marlowe tells her, "You looked good back there." Vivian has to be good if she's going to be a worthy love interest for the protagonist. The writer also adds the flirtatious dialogue between Vivian and Marlowe. Marlowe asks Vivian, "What's wrong with you?" She replies, "Nothing you can't fix." It's always good to involve the main characters in a romantic situation.

SIMILARITIES WITH SOURCE MATERIALS

The similarities between a screenplay and its source material are often greater than the differences. Such is the case in the adaptation of *Stand by Me* from Stephen King's novel, *The Body*. Here, a similar theme and social issue make for a similar story. In both stories, the writer focuses on the bonding that takes place among the four boys as they grow into manhood—a journey symbolized by their search for the dead body. Most of the plot actions in the original story and the adapted screenplay are the same, except for a change in the characters who perform them.

This is a common occurrence in most screenplays; an action is retained from the source but it is associated with a different character. In the novel, Chris pulls the gun on the gang, but in the screenplay Gordie pulls the gun. Thus, the writer is able to show character growth for Gordie, the protagonist, by showing him taking the action of standing up to the gang during the climax of the movie. In *Jurassic Park*, the protagonist defeats the dinosaurs in the novel. In the movie, he gets some help from nature and his ally.

Although there are many changes a writer can make when adapting material, all sources require the same imaginative processes one uses in developing an original idea or a story based on an imaginative character or plot line. Even though a character or idea might be based on a real person or incident, the writer changes the character or situation to make the idea original. In the same manner, a writer who bases material on a novel, short story, or play also changes the material by using her imagination. The process is the same, even if the original source is different.

In the past, many producers disagreed with the idea that original works and adapted works were similar. Writers of so-called adapted material were paid less than writers of scripts that dealt with original ideas. Many producers felt that the process of adapting a work was easier because the writer already had material provided to him in the form of a character, a plot line, or a theme. In recent years, this attitude has changed; writers of adapted material make the same or even greater amounts of money for their scripts. Studios have begun to accept that the adapting process may involve even more work than writing an original play or screenplay, because the scriptwriter may have to change entire characters and story incidents to make the script work within a dramatic context. Sometimes this involves considerable time and effort.

Approaches

Once a writer decides to deal with an adaptation, there are two fundamental approaches to develop a screenplay. In the first one, the writer can seek to forestall criticism that his work is disappointing because it doesn't remain faithful to the original, by sticking to his source material as closely as possible. But using this approach eliminates the possibility of the writer putting his own mark on the material because of his awe for the original material.

In the second approach, a writer can take a looser attitude. She can approach the material as if it were her own idea and put a distinct mark on it. Because film is a totally different genre than the novel, some writers argue that an adaptation is basically like an original and should be created in a fashion that bears the stamp of its creator's personal vision. In my opinion, the successful adapter must take a position somewhere between these two extremes.

Writers must approach the material with a certain amount of reverence, but not enough to become totally absorbed in preserving the original writer's vision. This reverence is the biggest problem for students who are doing an adaptation for the first time. They feel a duty to preserve the original at the expense of their own work. On the other hand, a beginning writer should keep in mind that the idea she is working on is not really her own. She should feel some obligation to preserve the integrity of the original material, if possible, but not at the expense of the new work.

The adapter, therefore, walks a tightrope. His chief concern should always be to make the original material work within a tight dramatic structure. If he can preserve original characters and plot lines to accomplish this goal, then all the better. But a writer must be prepared to make changes so that the story works. Unfortunately, many beginning students use a novel, short story, or play as a crutch and are afraid to be too ruthless, even when changes are really necessary. Often such screenplays fail.

The reasons for changes are usually pretty obvious. The original story isn't dramatic enough as a novel, novella, or short story. The novel or short story is a narrative style that does not readily lend itself to dramatization. Sometimes novels and/or short stories can be very introspective, with the writer interpreting what the character thinks and why he acts the way he does. Of course, in the screenplay, with some comic exceptions (Woody Allen's *Annie Hall* gives us the thoughts of the characters after they first meet), the writer has no way to reveal his characters' thoughts except by externalizing them in dialogue. Occasionally, voice-overs in films work to reveal inner moments. Sometimes characters even talk to themselves, but, for the most part, the characters reveal themselves through their interactions with other characters.

Potential Problems

In addition to knowing which fundamental approach to use, there are other potential problems when dealing with the novel or short story. For a screenplay, the writer needs to reveal the emotions of his characters. Using emotion serves to hook the audience into your story. But sometimes novels and short stories, especially more modern experimental ones, don't reveal their characters' emotions. Instead, these stories talk about emotions. Without sharing emotion, however, the main character in the screenplay is never tested.

Without a test, the story doesn't generate suspense and the potential conflict between the characters is weak. This is especially true of stories that are philosophical. A dramatic story must contain characters whose emotions are being tested under some unusual circumstances and whose reactions to these circumstances cause the action to move in a rising fashion, generating a beginning, middle, and end to the story.

Short Stories

In a short story, the main character is usually not fully developed. The necessary motivation and information so abundant in the novel is absent. Again, the writer must develop this character, give her emotion, and find a way to reveal this emotion through dialogue and action. The creative process of adaptation often follows the same steps required in the creation of a screenplay based on an original story.

Frequently, the material of a short story provides no characters to which the main character can react. Thus, as with an original story, the writer must supply characters from his own imagination. In a short story, the plot may be too static. Again, as in the original, the writer must provide action. The novel has an abundance of backstory; however, in the short story, this backstory may be nonexistent. Thus, it is up to the writer to provide the missing information.

Depending on the source material, the adapter is forced to either cut and condense the original material, such as with the long novel or play, or to expand the existing material, as in the case of the short story. For example, the novel may contain too many extra scenes that slow the pace of the story. The novel may contain too many characters, subplots, or scenes that are too interior. These characters and scenes may need to be cut to make the story more visually suited to the film medium. The key here is to be judicious. Pick relatively static scenes and either cut them to their basics or eliminate them entirely if they don't advance the story line. On the other hand, don't totally disregard the original source material, especially in the case of classic stories in which even minor characters are well known.

Sometimes it is necessary to change the location of the scene. If too many scenes take place in the same place, as in a play, or too many scenes take place in interior locations, it may become visually necessary to open the work by taking scenes outside.

In the case of short stories, the writer must add material to make the script long enough to fit into the time requirements of a movie. This process might include the addition of scenes, characters, and plot lines, often from very sketchy source material. Even for shorter dramatizations for cable TV or public broadcasting, the writer will need to expand the material if his original source is a short story.

Novels

Time is another element to consider when adapting from the novel or short story. In the novel, there are three time periods—past, present, and future. The screenwriter must deal in the present and devise ways to reveal the past through flashbacks or dialogue between characters in the story. The novel deals with internal reality; the screenplay deals with external events. The rendition of internal states and memories cannot be easily represented in film. The stream-of-consciousness novel, so popular in the twentieth century, is difficult to adapt to film. Representations of dreams on the screen tend to fail even when the film uses dissolves and superimpositions.

Of course, other differences between a film and a literary work can also affect the adaptation process. Film is more subject to the appeals of a mass audience than books, which may have a more limited audience. The scriptwriter has to make her story commercial and worry about the rating the film will receive if it's to make a profit. Novelists don't have to worry about censorship of their material. There is no X rating for books.

Novels and short stories are singular works. Films are products of collaboration. The screenwriter has to write with the idea that a director and other people will transform his work to the screen. Note that in a play, the director cannot change any of the writer's words without permission. Unfortunately, this is not true in TV and film. Thus, the conflict between writing what he wants and the demands of others involved on the project often can weaken the screenplay. The novelist has a single editor to please and, even then, isn't subject to the changes that editor may seek to impose on her.

The length of a novel differs from the length of a film. A person can read a novel over days and months. Novels tend to be longer and use more involved plots, subplots, and mythical elements. A film needs to be told in one two-hour stretch, use fewer characters, and rarely is mythical; but it has the advantage of editing and visualization. Novels can evoke more from one's imagination because all the "pictures" are supplied by the imagination of the person reading the book.

SCOPE OF THE CHAPTERS

The following chapters explore the adaptation process in more detail. They focus on the problems one encounters with characterization, structure, social issues, themes, and dialogue when adapting from a novel, short story, or play into a screenplay. The initial chapters deal with general problems involved in developing a screenplay, such as characterization, structure, and social issues, with some case histories interspersed—the adaptation of *Kramer versus Kramer, The Bridges of Madison County, Jurassic Park, Leaving Las Vegas*, and *Apollo 13*. Even though many of the examples in the chapters involve choices writers make when adapting from a novel into a screenplay, they can be easily applied to the problems of adapting from a short story or play. The main point is that the writer of an original or the adapter of the novel, short story, or play all encounter similar problems of character, structure, and issue when developing a successful screenplay.

Although most adaptations come from novels or novellas, some source material is from short stories or plays. Thus, we look at the differences in adapting from these media into the screenplay as well. Later chapters deal with the specific problems one encounters with the short story and play genre. If one chooses a short story or play as the source material for an adaptation, he or she should still read the earlier chapters on characterization, structure, and social issues before approaching the later chapters.

USE OF EXERCISES

At the end of each chapter there are several exercises designed to give potential screenwriters practice in developing specific writing skills. (The bold numbers at the end of a section indicate which exercises are meant to be used with the material you've just read.) The instructor can also suggest specific exercises and/or assign specific readings to sharpen a particular skill. Sometimes it may be necessary for students to consult outside sources to compare a particular scene or story line. At other times, the instructor may want to supply handouts of specific scenes for comparison and discussion in class. Students working on their own certainly will want to consult some of the outside sources suggested in the exercises.

The writer of original material can easily profit from these exercises, as well, because so many of the problems one finds in adapting can also be found in working with original material; namely, the problem of making the material dramatic, commercial, and interesting.

MASS AUDIENCE

The problem of dealing with a mass audience is discussed throughout the book. This problem makes the adapting process especially difficult, because novels or short stories only have to reach thousands of people to make a profit, while a movie must reach millions. Also, the tastes of the audience watching the film are continually changing. Thus, a novel, short story, or play written in the 1960s has to be updated to appeal to the 1990s audience. There may be a certain political bias in a book due to the time period in which it was written. Such a bias has to be changed to reflect not only the new times but also the political taste of the writer of the screenplay.

The purpose of this book is to examine screenwriting from the perspective of the special problems a writer may encounter when writing an original or adapting another's work to today's audience. It takes a nuts-and-bolts approach to the writing process, rather than a theoretical or critical approach. Many books deal with the differences between film and novel, but this book looks at screenwriting from the perspective of the writer and the problems one can encounter during the writing process.

THE QUESTION OF RIGHTS

One potential problem a writer has to deal with when working to adapt material from other sources is the question of rights. If someone picks a famous novel, short story, or play with the intention of selling this work after adapting it into a screenplay, the rights to this work must be obtained. One solution to this problem is to pick a work that was published several or many years ago. It could be a work that has declining sales and, therefore, might be optioned for a lesser fee. It is probably a good idea to avoid current best-sellers for which the cost of obtaining an option could be prohibitive. Of course, if a work is older or more of a classic that has fallen into the public domain (determined by the number of years since the author's death), then the writer doesn't have to worry about the cost of obtaining an option to use it.

In some cases, it is more advantageous to pick the work of a lesser-known writer, someone who might be more willing to option it for a lesser amount or even agree to share in the potential profits of a screenplay based on her work. When picking source material, a potential writer would do well to stay away from the works of famous writers such as Faulkner, Hemingway, or Neil Simon. Then again, the writer might look at this book not so much as a chance to sell a completed screenplay, but as a chance to sharpen one's writing skills, which might be of some future benefit. In this case, the choice of source material might be more flexible, based more on stories with dramatic potential and interest to the writer. In my classes, students write 60-page adaptations, instead of full-length screenplays, as an exercise in developing their writing skills. If the initial adaptation is successful, students can attempt to obtain the rights to the work and expand it into the longer 120-page screenplay.

SUMMARY

The problems of adapting material can be imposing, but, as with the writing of any dramatic material, success comes from the writing process itself. No text can teach anyone to be a screenwriter, unless the desire comes from within. Once a person has this desire and a few good ideas, the next step is to learn the craft of writing—this can only be accomplished by the continual process of writing.

I hope this book points out some of the pitfalls that lie ahead. The exercises and examples in the following chapters are only a starting point. The potential writer must sharpen her or his skills by actually writing. Even though this can be painful, it can also be rewarding. The journey may lead from learning a polished craft to perfecting an exciting art form. A well-written screenplay is as much an art form as a painting, novel, or piece of sculpture. And the well-written movie can be watched over and over again, revealing new insights into characters and the issues of the story. The next chapter looks at the elements that serve to transform any story into a dramatic one.

2

Principles of Dramatic Writing

It is essential for the potential scriptwriter to understand the principles of dramatic writing. Each subject has its own vocabulary; the same can be said for scriptwriting (see Terms to Know at the end of this chapter). Therefore, it is important to understand this terminology as we look at the adapting process.

COMMERCIAL POSSIBILITIES AND IDEAS

Every script begins with a basic idea. Whether the idea is original or comes from a novel or short story, it must have the potential for being dramatic. But what is drama? The clash of powerful ideas? The conflict between two powerful characters in a story? The rising suspense of a story that keeps you on the edge of your seat and surprises you with an unusual ending? The story of a person gripped with problems and obstacles trying to accomplish a basic goal? Is drama comedy, tragedy, action, or adventure, or a combination of all these elements? Maybe the successful script tries to incorporate as many dramatic elements as possible into the structure of its story.

Scriptwriting begins with an idea that must have the potential for being dramatic. This idea should be original, unusual, and perhaps commercial (somebody wants to buy it). Are you already saying that there are no new ideas—all the great plot lines have already been used in movies? Well, then you must find something different in a short story or novel, and some new way to make your script stand out from many others in the marketplace.

Maybe you have an interesting character you plan to explore. Maybe you're examining a new issue or new twist to an old story line. Maybe you've found some new way to update an idea by relating it to a current social problem. These features are key to making any idea commercial and marketable.

SOURCES FOR IDEAS

The newspaper is one of the greatest sources for new ideas and original stories, especially for TV movies. For example, a few years ago a writer was looking through the newspaper for story ideas. She found an article about a woman who had been raped. Considering the number of rape stories that have become movies or TV movies, this idea was not that unusual. But there was something different about the story. The rape had occurred in an elementary school. Was it one of the students? Was it a teacher or administrator? This idea was radical. The writer contacted the teacher and signed up the rights to her story. She then turned around and sold the story to one of the networks. A year later it became a TV movie starring Patty Duke. The story was different. It was commercial. It had a good social issue, namely, the lack of security on elementary schoolgrounds.

A few years later, this same writer was doing research using magazine articles. She came across an article about *Playboy* bunnies. Once again, this was not particularly different or fresh. Stories had already been done about *Playboy* bunnies. But this story was different. It was about Gloria Steinem, the well-known feminist. She had gotten herself a job with *Playboy* to expose how degrading it is for women to work for the magazine. This idea had commercial possibilities. In a year's time, this story also became a TV movie.

I'm not trying to suggest that every unusual idea has commercial possibilities. It takes time and patience to find a story that is different and appeals to a large audience. A student in my advanced writing class recently was working on a book that deals with a family relationship complicated by a terrorist kidnapping. The student's first impulse was to follow the story line of the novel. This is commendable in that writers should try to be as faithful as possible to their original source material. But sometimes this impulse is not possible. Stories about Arab terrorists have already been done. This story needs something new and different. It already has a relationship angle, which is good, but it needs to be updated and brought into line with the present political situation. The story would be better if it could anticipate some future event, such as the commercially successful film *The China Syndrome* that anticipated the Three Mile Island disaster.

Good art mirrors life. In the case of the student's story, I suggested that he change the terrorists to some group that hasn't been written about that might emerge in the next few years. The student wanted to switch to black militants or South American terrorists; however, stories involving these groups have already been done. Finally the student came up with the idea of making the terrorists Americans. This is a different approach. Radical American groups are becoming more prevalent in the news. But are they likable and sympathetic enough? By making their motivation government corruption and government cover-up, they gain some sympathy. Now the story has more possibilities. There's something marketable about it and, if indeed such a group does become involved in violent actions, the story will have even greater commercial possibilities. In fact, the story of the bombing of the federal building in Oklahoma is probably going into production even now.

STRUCTURE

Once the writer has an idea, of course, the next step is to put it into some type of order. One key to successful dramatization is structure. There are, however, many differences between the structure of a novel or short story and the structure of a screenplay. In the first place, the starting point of a screenplay is crucial to grabbing the audience's attention and also revealing the protagonist's characteristics and conflicts. This starting point should place the story's protagonist in some troublesome situation that escalates into a full-blown problem by the end of Act One.

Basically, all dramatic pieces follow a three-act structure, which corresponds to the beginning, middle, and end of a story. In the first act, also known as the set-up, we meet the protagonist, the antagonist, and other supporting characters who may be part of the main plot or part of the subplot—a parallel story involving minor characters that adds dimension to the main plot. The first act also establishes the conflicts of the story. With contrasting characterization, we might cause internal conflict within the protagonist. With the introduction of the antagonist, we set up additional conflict between two characters following opposite goals. We may also have conflict in the form of the protagonist versus nature or the environment. In *Jaws*, for example, the protagonist's internal conflict revolves around his fear of the water. He has to force himself out on the boat after the shark; he also fights against the character of Quint, the shark hunter, because Quint has radical methods for killing the shark. And, of course, he is also fighting nature (symbolized by the shark).

Act One ends with the protagonist having a definite problem she can't solve. Act Two opens with the protagonist temporarily solving her problem, but further believable complications or additional obstacles to the protagonist's goal arise and cause the problem of Act One to quickly escalate into the crisis scene at the end of Act Two. Here, the protagonist's problem has intensified into an extremely dangerous situation.

Act Two further develops the subplot of the story, showing us more conflict among the various characters. This act further develops different sides to the social issue, the topic on which the story is built—rape, wife-beating, drugs, abuse of political power, abortion, violence, gangs, incest, homeless families, single parents, animal rights, pollution, immigration, religion, and so on.

In Act Three, the protagonist once again solves her problem but fate, or an additional obstacle, quickly enters to cause the problem to reappear. We now build to the most dramatic scene in the story, the climax. Will the protagonist succeed in solving her problem or will she succumb and be crushed and destroyed? Members of the audience wait on the edge of their seats. We call this suspense, until a surprise twist at the end of the film occurs allowing the protagonist to solve her problem for the final time.

What follows is the ending—either positive, negative, or ambiguous. The writer chooses the best ending for his story, but audiences prefer to leave the theater with a happy ending. Of course, the ambiguous ending seems more artistic and does allow for a possible sequel. At the time of this writing, the happy ending is still in vogue.

The writer also reveals his theme or point of view on the social issue of the movie through the ending. Note that the writer never comes out and tells us the point of a story. He is more subtle; through the resolution or ending, the writer indirectly tells us his view on the story's issue. Character action reveals the author's theme. **[Exercises 1–3]**

SYMPATHETIC CHARACTERS

Once the writer has possibilities for a plot line, the next step is to create interesting, sympathetic, and well-rounded characters. The characters must have dimensionality and be complex enough so that they have contrasting characteristics—characteristics that create the potential for internal conflict. Let's say we have a character who is petite, well-dressed, 65 years old, a schoolteacher, and very prim and proper; when she leaves school to go home, she puts on a black leather jacket and hops on a motorcycle. This is contrasting characterization. On the one hand, the character has a proper facade; on the other hand, she obviously has a wild side to her personality. Like most people who are complex, she can act one way one moment and be an entirely different person a moment later.

Backstory

To develop interesting characters, I ask my students to write backstories and character sketches for each character in their stories. The purpose is to lay out detail that can later be eliminated or incorporated into the story. It's important to know your characters as well as you know yourself. What does the character look like? Any interesting tags, physical elements that symbolize the inner being of the character? *Miami Vice*'s Sonny Crockett had his flashy car. Kojak had his Tootsie-roll pops. Is there anything that can be a contrasting characteristic on the physical level for your character? Columbo, for example, dressed like a slob, drove around in a beat-up old car, and wore old clothes; on the other hand, he was very meticulous when it came to solving crimes. He had two very different character traits that defined his personality.

After you deal with the physical aspects of your characters, including the actual casting that forces you to visualize the character, it's time to deal with the personality of the character. Where does the character live? Who does he live with? What are his relationships like? If he's a father, is he a father like the one in *Father Knows Best,* the type of father in *Kramer versus Kramer,* or the military father of *The Great Santini*? Each is very different.

What is the character really like? Is she shy, an extrovert, only has a few friends, has an inferiority complex? What has happened to the character in the past to make her the way she is in the present? What is her ghost? She went to a preschool where she was molested and now she doesn't want to have any kids. She's afraid to marry your protagonist because of an incident that happened in her past. Of course, this incident occurred before your story began, but it explains her motivation.

On the other hand, maybe you have a character who comes from a divorced family. As a result, he has developed a negative view about the stability of marriage. That's why he's a playboy in your story, willing to give up the one person whom he really loves and who loves him. What is the character's goal in your story? Once you have a goal you can place obstacles in the character's path to prevent him from achieving the goal.

Character Growth

It's very important for the main character, the story's protagonist, to experience internal change from the beginning to the end of your story. If there is no change, then the character hasn't learned anything from the obstacles in his path. Thus, there is no theme, or point, to your story. In *Rocky*, the story begins with Rocky Balboa as a broken-down fighter working for a loan shark. During the course of the movie, he develops a goal—to become heavyweight champion. Something is at stake for the character—his image. All Rocky wants to do is go a few rounds with the champ so he won't look bad. He trains, becomes involved in a real relationship, and gives up his job as a collector for the loan shark. At the end, he doesn't win the fight, but he has changed from the loner we see at the beginning of the film to the well-conditioned family man who realizes his relationship with Adrian is more important than his image in the ring. Rocky has shown character growth.

Rounded Characters

It's not enough to create sympathetic main characters the audience can root for during the course of your story. Your characters must also have depth. If the character is too good, then there is no depth and no room for growth. If the character is too negative, then the audience can't root for him during the course of the story. Thus, it's important, through personality, physical traits, or situations, to make the character likable, but not perfect. If the character is a killer or rapist, then the writer must at least find positive qualities about the character for the audience to identify with and to make his crime and situation understandable. This, at least, builds audience empathy.

If the protagonist needs to be rounded so that she has more dimensionality, more interest to us, and more potential for internal conflict, it is also interesting to round out the antagonist of the story. In other words, one should make the "bad guy" not so bad and the "good guy" not so good. The most interesting stories have dual protagonists. These are stories in which both characters are sympathetic but, by pursuing opposite goals, generate conflict and problems for each other. The audience that is caught between two likable characters often roots for one and then the other. This is the case in the movie *When Harry Met Sally* in which both characters were sympathetic and acted as antagonists for each other. The audience was caught liking and disliking both characters.

If the story has a clear-cut antagonist, then it is necessary to make your antagonist more complex, more human, more likable. This is the case in the movie *The Paper Chase* (Bridges, 1972) in which the antagonist, Kingsfield, is rounded out several times by the writer James Bridges. We see Kingsfield, for example, in one scene arriving to teach his class hours before the students. Like the students, he is human. He must prepare for his class. The fact that he commends Hart for calling him a "son of a bitch" again shows that the character is human. He has a sense of humor. Through specific scenes and the reactions of other characters (we always see Hart admiring Kingsfield), the potential writer can find ways to make her antagonist more likable and more well rounded. **[Exercises 4–9]**

SUMMARY

The purpose of this chapter has been to give a brief introduction to some of the necessary elements one needs to write a successful screenplay. So that later chapters can be understood more readily, terminology used throughout this book is defined and discussed in the Terms to Know section that follows. No matter what type of material the writer is working with, the successful screenplay should incorporate as many key dramatic elements as possible. Even stories with these elements can fail, but stories without the potential for dramatization are likely doomed from the very beginning. Other problems and solutions that exist for successful dramatization of one's ideas are the subjects of other chapters. Next, we will look at problems that can prevent the development of successful characters.

TERMS TO KNOW

Action between the dialogue describes the business the characters are doing in a scene. When there is adequate action in the scene, the scene does not stagnate. The action is also used to reveal characterization.

Connected dialogue refers to lines of dialogue that contain one or two key words from a previous line and act to connect one line to the next.

Comeback line refers to dialogue that acts as a retort to a previous line of dialogue from a character in the scene.

Dialogue for character revelation is used to reveal character traits and motives along with action in a scene.

Dramatic need refers to something the character desperately needs in the story, and what is at stake or in jeopardy if the character fails to solve a problem.

Exposition is information about the character provided by the writer. It can also be the backstory.

Flashback is a structural device the writer uses to take the audience back into a past time and reveal backstory. This is not recommended for use unless the writer has no other way of subtly revealing the character's background. If one uses flashbacks, then give us brief scenes as in the style of *The Silence of the Lambs*.

Foreshadowing can be used to suggest what will happen at the end of the story. In *Harold and Maude*, for example, Maude tells us early in the movie that she will be eighty next week, and it will all be over. She is referring to her future suicide. When it happens, we are not totally surprised. We have been prepared for it.

Ghost is backstory haunting the protagonist from before the beginning of the story.

Hook is an event that occurs in the beginning of the story to grab the audience's attention.

Intercut is a film term to describe cutting two scenes together such as a telephone conversation. This device allows the audience to see both sides of the conversation without just listening to a boring one-sided conversation.

Master scene script is a script that contains descriptions and dialogue, but very few camera angles. This represents the type of script most writers use to submit their material.

Physical tag is an object or character trait that symbolizes an inner quality of a character.

Play on words is a dialogue device in which the writer plays with the words from a previous line of dialogue to connect two lines in a clever, amusing fashion.

Plot point refers to a moment in the action of the story when something significant happens to turn the story in another direction.

Premise is a three-paragraph synopsis that tells the beginning, middle, and end of a story.

Preparation prepares us for the ending of the story. In *Jaws*, for example, the writer prepares us for the killing of the shark by calling our attention to the scuba tanks brought on board the boat. We see one of them break loose and roll down the deck. Then, at the end of the story when the protagonist blows the shark up by shooting out the tank, we are not surprised that he was able to find a tank aboard the ship. The writer has prepared us for the protagonist's solution to the problem.

Rising action refers to the structure of a story where the action or pace continually increases as the story progresses. This device generates suspense.

Scenes are the dramatic units of a screenplay and are determined by location and time. Every time the location changes, the scene changes. Every time the time changes, even in the same location, we have a new scene.

Shooting script is the script that is developed from the master scene script and includes every camera angle in the story. This script is usually prepared by the director, rather than the writer.

Tag line refers to the last or next-to-last line of dialogue in a scene in which the character wins an argument with an incredible line that cannot be topped by the other character.

Treatment is a longer version of the premise, giving all the elements of plot and introduction of characters and can even include backstory on characters.

Unity is the process by which a writer tries to tie various elements of the story together, such as the beginning and ending, by using similar scenes.

EXERCISES

1. Watch several movies and see if you can identify the protagonist's problem and goal. Can you identify the act breaks? In a two-hour movie, the first act ends approximately 25 minutes into the movie. The second act, which is longer, occurs approximately 90 minutes into the story.
2. Perform Exercise 1 for both a half-hour and an hour episodic TV show. Note that half-hour shows have two official acts and hour shows have four acts, but these acts are all artificial. All dramatic stories really follow a three-act structure. Can you identify the three-act structure in the stories you are watching? What about drama written for cable TV or PBS? Is the structure any different?
3. Create a believable protagonist. Give her a problem. What obstacles or events can you create to make the problem worse? What would the climax of your story be? Would there be a twist or unusual ending?
4. Take one of the following characters and describe them. Give their personality and backstory and list specific actions, situations, or tags you would use to make the character sympathetic. *Note:* The following characters are slightly stereotyped; find ways to make them more well rounded.

a. Heather Gottlieb
b. Marvin Brown
c. Freddy Hernandez
d. Todd Little

5. Pick several recent movies and identify the protagonist and antagonist. What are the characters' goals? What's at stake in the story? How does the writer round out the antagonist?

6. Write a scene that demonstrates conflict using two of the characters in Exercise 4.

7. For one of the characters in Exercise 4, write a long speech that reveals something interesting about the character's past.

8. Write a three-paragraph premise on a story idea in which you are interested. If the story has a beginning, middle, and end, fill in the details by writing a longer treatment.

9. Watch several contemporary movies and see if you can find moments of preparation and foreshadowing. List a moment of foreshadowing you would use in the treatment for the story you have just written for Exercise 8. Discuss the movies you watch in class.

3 Characterization

The necessity to create interesting, sympathetic, and unique characters for dramatic effect is complicated by the process of adapting the novel or short story into the screenplay. As noted critic Thomas Craven said (1926, pp. 489–90):

> I doubt if the most artistic and sympathetic reader ever visualizes a character; he responds to that part of a created figure which is also himself. But he does not actually see his hero. For this reason all illustrations are disappointing.

No successful screenplay can afford to be disappointing. Its characters must excite and intrigue us while commanding our emotional involvement. We must root for them, agonize with them, experience their internal conflicts, and share the joys of their triumphs as they move through the pit-filled landscapes of dramatic stories.

Yet the problem that remains is how to transpose a character from the written page to the screen, and still remain faithful to the original writer's creation. Can the screenwriter create exciting, well-rounded, visual duplications? Will cinematic characters always take second place to their original creations? Are adapted characters harder to bring to life than entirely original ones? The answer is no. If certain fundamental principles are followed, it is very possible to create three-dimensional, believable, and emotionally interesting characters. This chapter examines these principles, focusing on the following four key tools that can be applied to the transformation of a character from a novel to a screenplay:

1. *Create sympathetic or empathetic characters.* Overcome negative fictional heroes or heroines.
2. *Focus on relevant backstory.* Eliminate and narrow down the abundant background information presented by the original writer.
3. *Create precise and brief character descriptions.* Boil down two pages of character description into five to ten succinct descriptive words.
4. *Show character motivation.* Turn narrative explanation into visual action.

THE SYMPATHETIC PROTAGONIST

The first task in successful characterization is to make the audience identify with the main character. In the adapted screenplay, this task is further complicated by the very presence of the novel, short story, or play. The visualization of a character in a film removes part of the mystery that makes the character so likable in the original work. The difficulty of recreating the mystery and making the character interesting for the viewer is one of the prime aims of the scriptwriter. He accomplishes the task by creating a three-dimensional representation of the character.

Identification

Besides likability, however, the protagonist must also be a universal figure, struggling with the problems and obstacles of the story. This goal of universality can be greatly complicated when the audience watches Paul Newman or Tom Cruise portraying a favorite fictional character. The audience has a hard time not confusing your hero with the mystique of these well-known actors. The scriptwriter can help herself, however, by creating such believable characters that the audience can immediately step over the hurdle of the personality of the well-known actor. This effort toward believability can help overcome the negativity of substituting a movie character for a well-known fictional one.

Rooting for the Protagonist

In addition to identification, the audience needs to root for the protagonist as she or he confronts the problems of the story. The scriptwriter achieves this task by creating sympathy or empathy for her protagonist. Often, the scriptwriter needs to change the original characterization to achieve this aim. In *Wuthering Heights*, for example, the novelist's depiction of the protagonist, Heathcliff, is often negative. As Bluestone

(1957) described him in *Novels into Films*, he is more demonic, more negative, and more violent with his "diabolic sneer, eyes like devil spies," and "sharp cannibal teeth" (p. 107).

The scriptwriter needs a sympathetic protagonist. He creates one by turning Heathcliff's violence into romantic passion. He makes the protagonist's violence originate from the pain of a man deeply in love, not the pleasure of a man who enjoys hurting and abusing people. This change in temperament appealed to the romantic tastes of the audiences of the time. Audiences can root for a man tormented by love, because they can understand and identify with him. Everyone has had painful love experiences. On the other hand, no one wants to be reminded of his potential for violence.

Changes in Situation

Often sympathy is created for a character by minor changes in the protagonist's current situation. One element screenwriters use to reveal a character is vocation. In *Heartburn*, for example, Nora Ephron makes her novel's hero a famous TV personality. Yet the screenwriter made her into a cookbook writer and minor magazine journalist. This apparent unimportant change, however, makes the protagonist more financially dependent on her husband. When the relationship with her husband is threatened by his infidelity, she is in a more precarious situation. Her increased vulnerability in the screenplay makes her more sympathetic to the audience.

Likewise, in *Kramer versus Kramer*, the vocation of the protagonist is changed to evoke greater sympathy. In the novel (Corman, 1977), Ted Kramer sells advertising space in various magazines. The screenwriter intensifies his job—he becomes an account executive for a major advertising agency faced with the daily pressures to produce ad campaigns, meet with clients, manage problems, and deal with competition. As the protagonist's situation intensifies, sympathy from the audience also increases as he is caught in a mounting conflict between keeping his job and taking care of his son. Ultimately, when he commits more time to his son, Ted loses his job. This, in turn, leads to further complications when his wife Joanna sues him for custody of their son. Without a job he has no chance of winning custody. He's desperate, vulnerable, and scared. Now the audience is really hooked.

In *Jurassic Park*, the little girl is modified from the novel (Crichton, 1990) to make her more sympathetic and more of an ally for the protagonist, Grant. In the novel, she is fearful and weak. In the movie, she becomes more computer literate, less helpless, and more appealing to modern audiences by being a stronger woman. Ellie, Grant's girlfriend, has also been modified. She is stronger and given to more speeches on women's issues in the movie than in the book.

Changing Other Characters

Sometimes a writer achieves character sympathy not by changing the protagonist, but by changing the other characters that surround that person. Remember, often a writer reveals a character by what other characters in the screenplay have to say about him or how other characters react to him. In Raymond Chandler's *The Big Sleep*, the protagonist, Marlowe (played by Humphrey Bogart), has an affair with Vivian Regan, a married woman. In the screenplay, Vivian becomes Vivian Rutledge, the ex-wife of Sean Regan. Thus, Bogart doesn't have an affair with a married woman, an event sure to lose him sympathy with the conservative audiences of the time. He has integrity.

In Chandler's novel (1975), *Omnibus*, Marlowe is a womanizer. He thinks an affair with Vivian might make a "jazzy weekend but wearing for a steady diet" (p. 77). In the movie, Marlowe falls in love with her. Overall, the writer enhances Marlowe's character in the screenplay by the changes made to minor characters. He is made more attractive to women. He flirts with the vampy Dorothy Malone, the bookstore clerk, and also attracts the attention of the cute taxi driver who is taking him around. In the novel, the bookstore clerk is described as curt and businesslike, while the taxi driver is more preoccupied with her horror magazine than with men. Thus, through the use of other characters, our respect and admiration for Marlowe grows. Attractive women like him. We like him.

Another example of minor character changes that affect the protagonist's sympathy can be seen in the adaptation of the novel *Heartburn*. In the novel, Rachael's confidant, Richard, is both her lover and TV producer. In the film, however, he's just Rachael's confidant, not her lover. By making the protagonist's relationship with him a nonsexual one, the protagonist gains more sympathy when her husband is unfaithful. His unfaithfulness is not caused by the protagonist. We can identify and feel sorry for a wronged wife, but not an unfaithful one.

Sometimes a protagonist gains sympathy by reducing the importance of another character in the scene. In *Marvin's Room*, for example, the novel (McPherson, 1992) places equal emphasis on Dr. Wally and Bessie in the initial scene in the doctor's examining room. Dr. Wally is made humorous because of his speech about bugs in the office. Bugs he tells us must be in her imagination since they are living in Florida. At the next moment, a bug walks by. It is a humorous moment that is cut in the screenplay because it puts more emphasis on Dr. Wally in the scene and not on the protagonist. In the movie's scene, the audience needs to feel sympathy for Bessie because of her life-threatening disease, not laugh at Dr. Wally.

Changing the Dialogue

Sometimes merely changing the tone or speeches of characters can create more sympathy. In Webb's 1963 novel, *The Graduate*, for example, Benjamin, the protagonist, is very rude and curt with his parents and their friends. His parents appear to be more sympathetic and understanding of his plight. "And let's be honest about this, Ben, your mother and I are certainly as much to blame as you are for whatever is happening." Ben replies sarcastically, "This is getting melodramatic. . . . It has a kind of hearts and flowers ring to it" (Webb, 1963, p. 69). In the novel, Ben is sarcastic and nasty. In the film, the writer changes the dialogue and the tone of the two characters to make Ben the sympathetic figure besieged by his obnoxious, overbearing, and materialistic parents. He's the protagonist. We need to quickly like him, or at least to understand him. Unlike the experience of reading a novel, we don't have the time to gradually get to know and like the main character. **[Exercises 1–2]**

BACKSTORY

One of the key differences between the source material and a screenplay involves length. The short story can run only a few pages. A play can have long scenes. The novel can run 400 to 500 pages and even longer. The screenplay runs around 120 pages or two hours in length—one page equals one minute of screen time. One advantage a novelist has, therefore, is more time to explore the background and history of the characters.

The screenwriter has to be more selective. He has to decide what background information is irrelevant and what isn't; which will reveal character motivation and what information contains more detail than we need to know; what backstory information advances and maintains the tension of the story and what causes it to come to a grinding halt. If the writer is adapting from a short story of only a few pages, he may have to invent backstory for the characters.

One of the most popular ways of deciding what backstory to use in a novel or long short story relates to the actual starting point of the screenplay. In *Kramer versus Kramer*, for example, the first 100 pages are devoted to the backstory on Ted and Joanna: how they met, trips to Fire Island, the singles' scene, first date, their single friends. It's not until they've been married for several years and have had their child that we reach the starting point for the screenplay.

Structurally, because a screenplay should begin with a troublesome situation that escalates into a problem, *Kramer versus Kramer* (Benton, 1978) begins with Joanna moving out on Ted. At this point, how Ted and Joanna met is totally irrelevant to the forward motion of the story, which then moves into Ted coping with raising Billy. The only background information the screenwriter gives the audience involves a scene that shows Ted's involvement with his job. Thus, Joanna's motivation for moving out is partially explained by Ted's workaholic behavior, creating sympathy for Joanna, at the same time that it gives Ted's character a starting point that can be changed later. We call this evolution character growth. Ted moves from a workaholic, inept father to a caring, capable one.

In *The Shawshank Redemption,* a large portion of the backstory about the protagonist's conviction for murder is eliminated. We lose the investigation, the trial scene, and even the information on how the protagonist settled his affairs before coming to prison. Backstory information on the ally is also eliminated. As Stephen King describes him: "I was young, good-looking and from the poor side of town. I knocked up a pretty, sulky headstrong girl who lived in one of the fine old houses on Carbine Street. Anyway it's not me I want to tell you about" (1982, p. 16). In the movie, the focus is on Andy, not Red.

In *Marvin's Room,* backstory not related to the protagonist is also eliminated. All the history of Charlotte forcing herself through school . . . how she kept a clean house and her relationship with Hank are cut. Backstory on Lee and her relationship with Hank is also omitted. Backstory on other characters can be revealed in screenplays as long as the information remains short and the emphasis remains on the protagonist.

In *The Firm,* backstory on the subplot character of Kay Quinn is cut. In the novel (Grisham, 1991), we learn of her small-town background, her marriage to Lamar after college, the trip to Vanderbilt law school, two babies in fourteen months, and her work with the garden club. Backstory on other minor characters, such as Oliver Lambert, is also cut. In the movie, we know about his marriage problems but not about his two kids, his million dollars in the bank, and his relationship with his wife. Exposition on the protagonist's relationship with his in-laws is also cut (Grisham, 1991, p. 15):

> I would be closer to your parents and that worries me. . . . She deflected this as she did most of his comments about her family . . . he imagined her parents' first visit . . . they would burn with envy and wonder how the poor kid with no family and no status could afford all this.

Exposition, which tends to interpret a character's feelings, is usually cut in the movie for a more visual presentation of the material. In *The Silence of the Lambs,* all relevant backstory on Crawford and his relationship with his dead wife has been cut. The emphasis in the movie remains on Clarice and her ghost, the death of her father. In the novel, Clarice interprets Crawford for us (Harris, 1988, p. 2):

> Clearly something was wrong with him. There was a peculiar cleverness in Crawford aside from his intelligence and Starling had first noticed his color sense and the textures of his clothing even within the FBI clone standards of agents' dress. Now he was neat but drab, as though he was molting.

In the movie, we learn about Crawford visually through Starling's reactions to him.

Starting Point

Often, the starting point for the screenplay and the source material are remarkably similar. In the adaptation of Shirley Jackson's short story, "The Lottery" (Angus, 1985), both the short story and the film begin with the children assembling in the town square. In the adaptation of Neil Simon's play "Barefoot in the Park" (Simon, 1980), both play and screenplay begin with Corrie entering the small apartment in New York. The short story writer, novelist, and often the playwright, however, are free to slow down and explain character behavior.

In D. H. Lawrence's short story "The Rocking-Horse Winner," Lawrence goes into great detail explaining the mother's motivation for appearing to be a good mother. "Yet what it was that she must cover up she never knew. Nevertheless, when her children were present, she always felt the centre of her heart go hard" (Angus, 1985, p. 91).

The screenwriter doesn't have the time to stop and explain the characters' behavior. She must be very selective in using only backstory necessary to establish some understanding of her characters. This is certainly the case in *Ordinary People*. In the novel, we learn, in detail, the history of Calvin—how he grew up in an orphanage, how his mother would come and visit him, how he met his mentor who put him through law school, how he met Beth playing tennis. The screenplay's writer doesn't include any of this information, because the focus of the story is on Conrad and not on Calvin.

Conrad's backstory in the novel is also extensive. We learn how he got straight A's in grade school and junior high, how he rode his two-wheeler sixteen times around the block on his sixteenth birthday, how he got four firsts in the 100-meter freestyle. In the screenplay, all these interesting details are omitted; there is no time for it. Instead, the only backstory we get on Conrad is subtle references to his attempted suicide and the death of his brother, Buck. Again, the writer uses backstory in the screenplay only when it is necessary to explain the actions of the character in the present situation and to create a troublesome starting point that will escalate into the character's problem by the end of the first act.

In William Styron's *Sophie's Choice*, it takes us 31 pages to get to know the character of Stingo. We learn about the origin of his nickname, his job as a manuscript reader (including his evaluations of various works), his life at the University Residence Club, and his relationship with his father. The screenwriter omits all that information. Instead, he tells us about the character's hopes, his goal of becoming a writer, the fact that apartments are not cheap in New York (problem), and that we are in Brooklyn in 1947 (setting and time).

The novelist tends to give us backstory in big batches. The screenwriter gives us bits and pieces, careful not to stop the movement of the story, yet keeping us interested in his character as we move along. Again, *Sophie's Choice* illustrates this point. In the novel, Stingo meets Sophie, sees her tatoo, and asks her what it is. In the screenplay, Stingo meets Sophie, but doesn't see the tatoo until a later scene. He doesn't ask her what it is, but we see him noticing it. The writer uses the visual imagery rather than language to develop the character's backstory. It's much more effective. **[Exercises 4–7]**

CHARACTER DESCRIPTIONS

Along with economy of backstory, the screenwriter must prune the lengthy descriptive passages of a novel or long short story down to a few key words. Take, for example, the description of Conrad in *Ordinary People*. In Guest's novel (1976), his "face, chalk-white is plagued with a weird constantly erupting rash. This is not acne they assured him. . . . He tries to be patient as his hair grows out. He had hacked it up, badly cutting it himself the week before he left" (p. 3). In Sargent's screenplay (1979), this becomes "CONRAD, 17, with a strangely chopped haircut just beginning to lay flat" (p. 1). Here, the screenwriter focuses on the key visual element of Conrad's unusual hair as the screenplay begins, giving us the impression that something is wrong.

Another character in the book is Beth. The novelist (Guest, 1976, p. 6) describes her

> . . . as having a face that is soft in the morning, flushed, slightly rounded, ycunger than her 39 years. Her stomach is flat, almost as

> if she never had the babies. . . . Beautiful hair, the color of maple sugar or honey. Natural, too. The blue silk robe outlines her slender hips, her breasts . . . all elegance and self-possession, so beautiful in every detail . . . a determined set to her chin that moves him even when she uses it against him.

In the screenplay, this becomes "BETH, 39, has the classic bone structure of the well-groomed upper strata American family" (Sargent, 1979, p. 2). Here, the screenwriter paints a clipped, precise image emphasizing not the hair or breasts, but the well-groomed, upper-class background. This is the basis of the character's problem and conflict. This upper-class perfection is in conflict with the violence and disorder of her son's mental breakdown. **[Exercises 8–10]**

CHARACTER ACTION

A major difference between the fiction writer and the screenwriter lies in revelation of character. The novelist, relying on the linguistic component of words, is able to reveal her character mainly through narration, but also through dialogue and reactions. The scriptwriter must use action, dialogue, and other characters to reveal his characters. Perhaps the most difficult problem for adapters is to translate the novel's narration into character action. Most beginning writers fail, ending up with screenplays that seem more novelistic than filmic.

In *Ordinary People*, Judith Guest (1976, p. 24) tells us about Beth:

> . . . self-possessed is what she is; he emphatically does not own her, nor can he understand or even predict with reliability her moods, her attitudes. She is a marvelous mystery to him. . . . Everything had to be perfect. Never mind the impossible hardship it worked on them all; never mind the utter lack of meaning in such perfection.

In the screenplay, the writer focuses on Beth's self-possession and perfectionism. Instead of telling us about them, we see them through particular actions in the script. In Sargent's second draft of the screenplay, we see Beth in the kitchen: "Beth moves into frame and she slides a saran-covered bowl of fruit salad into its place . . . neat as a pin" (Sargent, 1979, p. 2). In the next scene,

> Beth moves to a pad of paper, makes a note, then she moves to a chopping block and picks up a pile of neatly folded towels. . . . Her moves are graceful, controlled as she moves to a small antique desk. . . . She takes a pretty pencil out of a box of pretty pencils and makes a note on a pad . . . it reads, "Add Thompsons to list."

The screenwriter modifies the character. Her perfection becomes self-control. This self-control makes her unemotional toward her son and the source of much of the conflict in the screenplay. The relationship between Beth and Conrad is a constant source of conflict and friction in both the screenplay and the novel.

In the novel, we learn that Beth thinks Cal is too concerned with Conrad's problem: "She thinks you're obsessed with Con's problem. You can't think of anything else" (Guest, 1976, p. 149). In the screenplay, however, the writer dramatizes this difference in a breakfast shared by Beth, Cal, and Conrad. We see something is wrong in Beth's relationship with Conrad when she gives him French toast. She gets mad when he doesn't eat it and then abruptly takes it away.

```
                         BETH
          If you're not hungry . . . you're not
          hungry.

She takes the plate and moves to the sink with it.

                         CAL
          Wait a minute. He'll eat it. Con, it's
          French toast. (p. 12)
```

Beth is unemotional. Calvin is too concerned, but caring. In the novel, we are told about it; in the film, we see it.

Contrasting Characters

The writer also uses action to contrast his characters. In juxtaposed scenes, we see Beth in the bathroom "gently leaning over her marble sink . . . she takes her jar of cream, her fingers are in, to take just a small amount carefully and delicately rubs it in" (Sargent, 1979, p. 20). In the following scene, we cut to Calvin in his den, drink in hand, watching TV: "He reaches for his gut. Grabs it. Squeezes the excess" (p. 21). The contrast is obvious. The careful, delicate Beth, the down-to-earth, practical Calvin. He sees Conrad's problem; she ignores it. **[Exercises 11–12]**

SUMMARY

The key to adapting successful screen characters lies in making them sympathetic, narrowing down their extensive backstory, describing them in a precise manner, and using action to reveal them in a visual manner. In addition, the successful adapter should be concerned with other issues, which we discuss in subsequent chapters. Character growth involves the change a screenplay's character undergoes as a result of the story's problem. It may be necessary, therefore, to change some aspect of the novel's characterization of the protagonist to create a new starting point, which the character can evolve and change from.

Often, this change in starting point of the story can be achieved by changing the character's goal. Finally, the successful adapter should be aware of the minor characters in the story. By combining characters and eliminating others, the writer can achieve greater economy. The story moves quicker and holds the attention of its audience better. Successful characterization can never guarantee a commercial success, but it's an important first step. In the next chapters, we explore other important screenwriting steps—structure, social issue, and dialogue.

EXERCISES

1. Rewrite the two speeches of Benjamin and his father from *The Graduate* to make Ben appear more sympathetic.

2. In Ann Beattie's 1976 novel, *Chilly Scenes of Winter*, Charles, the protagonist, is in love with a married woman. Charles, a dull, civil servant, chases Laura through the course of the book. In the film adaptation, *Head over Heels*, the writer uses two specific actions to make this dull character more interesting. In one scene, we see Charles hard at work at his job writing reports on reports, but "he's been promoted twice." How can the writer make Charles less of a stuffed shirt, while showing he doesn't take his job totally seriously?

The screenwriter accomplishes this by giving the character a sense of humor. For example, when we see Charles working on his reports, the writer tags the character by giving him a cassette player and Janis Joplin music to rock out on. After he finishes a report, he opens his drawer and takes a drink. Here, a simple action makes him more human. We like this guy. He's obviously doing a good job because he's been promoted, but he's not a total bureaucrat. He's like the rest of us who like rock music and take an occasional drink. In the next scene, the writer, again using action, shows Charles as he leaves his office and walks by the secretarial pool. What does Charles do that shows he's got a sense of humor?

One incident involves an elderly secretary, the other one a black secretary. Make the action in your story brief and then turn to the appendix to see how the professional writer achieved his goal of making the character more likable.

3. How would you make a killer, rapist, or a gang member a sympathetic or empathetic protagonist? List specific character traits, character actions, and motivations. Check the appendix when you finish to see how this was accomplished in the adaptation of *A Clockwork Orange.*

4. In *Ordinary People,* the novelist gives us this background information on Jeanine, Conrad's girlfriend. She's not a virgin; she once hung around with a bunch of wild kids who were into stealing. Her mother and father were divorced. She plans to go to college and study music. She's just moved to Chicago because of her mother's affair with another man. If you could pick one vital piece of information to give us about Jeanine, what would you pick and why? Jeanine reveals this information in a scene with Conrad when he's telling her about his attempted suicide. Write the dialogue for this scene and have your instructor check it.

5. Write your own autobiography. Pick only three relevant items to include in a dramatic story based on your life. Why did you pick these items? Write a scene in which these items are revealed in dialogue. Be subtle.

6. Rent a video of a popular movie. Listen for dialogue that reveals backstory. Why did the writer pick this information to reveal to the audience? Consult the novel on which the movie was based. What backstory was omitted and why?

7. *Character Ghosts*: Take two movies and describe two backstory events that haunt your character throughout each movie but aren't revealed to the audience until the second or third act.

8. The key to a good screenplay description lies in precision, selectivity, and brevity. Using a few key words, metaphors, and similes and a clipped style, rewrite the following character descriptions from *Ordinary People.* Compare your descriptions to the actual ones from the screenplay in the appendix.

Jeanine:

> In front of him are two sopranos, one blond and one redhead, whose hair hangs silk smooth and straight, almost to the middle of her back. No, not red; more of a peach color. The redhead has blue eyes, copper-colored freckles and a blue skirt. (pp. 19–20)

Calvin:

> Calvin Jarret, 41, U.S. citizen, tax attorney, husband, father, orphaned at the age of 11. He caught himself thinking about that lately, thinking of the Evangelical Home for Orphans and Old People. An odd kind of orphanage. Most of the kids had at least one living parent. He had moved there when he was four. . . . He was

> named Calvin for his dead uncle. Jarret had been his mother's maiden name. Responsibility, that is fatherhood. You cannot afford to miss any signs. (p. 7)

9. Describe your own bedroom. Remember the writer uses selectivity in picking objects that reveal the character of the person who lives in that room. What objects reveal your personality? Do the same exercise for your bathroom.
10. One aspect of good description is comparison. Describe yourself in terms of sound, sight, smell, and taste. Use similes and metaphors, and concrete adjectives. Be precise.
11. Using action and dialogue with double meaning, write a scene involving a PTA night. A mother brings her teenage daughter to her art teacher's class. Unknown to the daughter, the mother has been having an affair with the teacher. The affair has just ended. Use the mother's dialogue and actions in the art room to reveal her real feelings about this breakup. Check the appendix for a sample student script.
12. Using action, write a comedic scene in which the protagonist is trying to seduce the wife of his best friend. Check the scene from Woody Allen's *Play It Again Sam* at the end of this book to see what action and dialogue were used to undercut the seduction and create comedy.

4

Structure

The most important factor in the adaptation process after characterization is structure—the organization of plot into a specific sequence of scenes. Structure is the key ingredient of a well-written screenplay. The rambling and loose structure of the novel must be tightened into the three-act structure of the screenplay, increasing the protagonist's problem until the climax is reached in the third act.

To achieve this increased pace, develop the characters, and tell the story, the writer must find key scenes; pick a suitable starting point; add scenes to fill in plot details; and, in general, condense and combine characters and scenes. Whenever possible, dialogue should be condensed and visual elements brought into play.

Depending on the initial source material, the writer is faced with two basic choices. If she is adapting from a short novella or short story, she can add scenes or expand existing ones. If she is dealing with the structure of a large novel, the writer may need to eliminate scenes and condense existing ones in order to build to a more dramatic ending. In most cases, the writer needs to eliminate a certain number of existing scenes from a novel. The reason for this is simple. The longer novel's leisurely pace, multiple subplots, numerous scenes dealing with backstory, and character motivation only slow the screenplay down.

A fast-moving story depends on tight structure. This chapter examines structure by focusing on these six key tools that can be applied to the transformation of the structure of a novel or short story into the structure of a screenplay:

1. *Find the starting point.* Pick the appropriate starting point for the screenplay.

2. *Eliminate scenes.* Decide which scenes to omit from the screenplay.
3. *Condense scenes.* Shorten the usually lengthy novel or play scene into the shorter screenplay scene.
4. *Combine characters.* Combine minor characters into one screenplay character.
5. *Add scenes.* Introduce additional scenes to fill in plot gaps.
6. *Rework beginnings and endings.* Change beginnings and endings to achieve unity.

STARTING POINT

Where does a writer begin a screenplay? Because of time constraints and the need to hold the audience's attention, the writer must follow a tight structure that starts with a troublesome beginning that grabs the audience's attention—for example, James Bond killing sixteen guys, sleeping with four different women, and speeding around in his latest sports car, shooting guns, and dodging bombs. Or the writer can begin with an opening that reveals the protagonist's problem, such as Warren Beatty's in *Shampoo*. The story opens in a darkened bedroom. We see Warren in bed with one woman and on the telephone with another one. This is his problem, as well as his conflict. *Leaving Las Vegas* opens with Ben shopping for liquor. This is his problem, which will worsen as Ben drinks more and more until he dies in the movie's third act.

Charlie Peters's (1980) *Paternity* opens with the protagonist, Buddy (Burt Reynolds), staring at kids playing in a schoolyard. We find out it's Buddy's birthday. This is his problem and conflict. He's 44 and doesn't have any kids. On his way to work Buddy meets one of his employees, Carlos:

```
                    BUDDY
        Carlos, how's the wife?

                    CARLOS
              (handing a cigar to Buddy)
        Eight-and-one-half-pound girl. (pp. 2-3)
```

This is Buddy's problem: He wants a kid, but not a wife. In the first or second scene of most screenplays, all the information about the protagonist's situation is revealed to us. For example, take the beginning of Waldo Salt's *Midnight Cowboy* (1969), adapted from the novel by James Leo Herlihy. As the movie opens, we meet the protagonist, Joe Buck (Jon Voight):

```
INT. TEXAS MOTEL ROOM—DAY

On the dresser—a counter display of male cosmetics in
their merchandising packages—a six pack of gum—a carton of
cigarettes—a fragment of chest and chin glimpsed in the
mirror as Joe sprays underarm deodorant—aims one playful
blast at his unseen crotch, another at . . . a blushing
calendar girl, advertising The Sunshine Cafe—her mouth
frozen in a tiny O, staring at Joe. (p. 1)
```

Here's the protagonist's conflict in a nutshell. He's caught between his own narcissistic loner fantasies and his desire to connect with women. The crotch versus the calendar girl. In the next scene the troublesome situation begins.

```
INT. CAFE SCULLERY—DAY

RALPH—an aging black man—scraping dirty dishes.

                         RALPH
              Where's that Joe Buck?

INT. TEXAS MOTEL ROOM—DAY

Joe's torso and chin, grinning as he tucks in his new cowboy
shirt and zips his tight Levis . . .

                         JOE
              Yeah, where's that Joe Buck? (pp. 1-2)
```

Here's the character's troublesome beginning and future problem. Note the connection of dialogue so common in the screenplay, but absent from the novel. Note how this connected dialogue tends to connect these two scenes while moving the story along. The writer connects two lines by repeating the first line. However, because the second line has a different emphasis, it isn't simple repetition.

Joe's employer wants him at work washing dishes. Joe sees himself differently: ". . . a poster of Paul Newman as Hombre watches Joe's feet slide into the boots" (pp. 1–2). He sees himself as a "Newman-like" stud. To satisfy this image he quits his job to become a hustler in New York. The starting point propels the story in a certain direction through the first act, when the problem intensifies and the story line takes another direction for Act Two.

Finding the Starting Point

The screenwriter must begin with a troublesome situation that will reveal the character's problem and conflict. Often, however, he has to work through pages of background information to reach this starting point, such as in *Kramer versus Kramer*—the writer has to go over 44 pages of background information to reach a suitable starting point for the screenplay. The screenwriter chooses to start his story with Joanna's leaving Ted. The novelist starts with Joanna giving birth to Billy, and Ted's guilt about his reaction to seeing her bleed. The novel's starting point makes Ted a caring husband who feels guilty that he can't help Joanna. The screenplay does not emphasize this facet of Joanna and Ted's relationship. The screenplay isn't about Ted and Joanna giving birth to Billy; it's about Ted coping with being a single parent.

The screenwriter needs a later starting point. By starting with Joanna's departure, the writer gives the story a troublesome beginning that leads to an exploration of the screenplay's central issue—which sex is the better parent? The writer begins by giving Ted a problem. How will he take care of Billy with Joanna gone? At the end of Act One, when it becomes apparent that Joanna is not coming back, Ted's problem worsens. Coping isn't a part-time situation—it's full-time—and could mean the loss of Ted's job and career.

The entire pace and rising action of the story depends on your beginning. In *Ordinary People*, the writer finds the starting point for the screenplay in the choir scene on page 19 in the third chapter of the book. In the novel, the story begins with Conrad in the mental hospital recovering from his attempted suicide. In the screenplay, the writer moves time along by beginning later and showing us Conrad back in the real world trying to cope with his school life. The story isn't about his suicide, but about how he tries to cope as an ordinary person in a not-so-ordinary world.

Sometimes writers have to dig deeper for their starting points. The novel *Out of Africa* begins with the protagonist in Africa talking about her farm, coffee growing, and the local natives. In the screenplay, the first several chapters of the novel are reduced to one page of script that dissolves into a room in a house in Denmark and the backstory of the protagonist, how she met her husband, agreed to marry him, and came to Africa. But this information is not in the novel.

The screenwriter searches deeper into Judith Thurman's biography of *Isak Dinesen* to expand his starting point. Later, in order to embellish on the relationship between Karen and Denys, which is never given as more than a friendship in the novel, the screenwriter consults another biography—Errol Trzebinski's *Silence Will Speak*, the biography of Denys Finch Hatton. In other words, the screenwriter often has to hunt for the interesting, relevant, starting material.

As Ben Brady (1974) says in his *Keys to Writing for Television and Film*: "It is necessary for the adapter to start the play with an exciting development that may not occur in the original story until you are well into it" (p. 246). **[Exercises 1–3]**

ELIMINATION OF SCENES

One of the three key choices the adapter has when it comes to dealing with the structure of his source material is to eliminate scenes. As Dwight Swain (1984) says, "The big issue is relative complexity. A novel may run 500 pages or more. Most films must be squeezed into 90 minutes running time. Cutting often, drastic cutting therefore becomes essential" (p. 187). Remember, 90 minutes equals 90 pages of script. How does the writer make this choice? Basically by eliminating unnecessary background scenes from the novel.

In the adaptation of *The Godfather*, all the scenes dealing with the young Vito, his coming to America, and his early days in New York are interesting, but unnecessary for dealing with the protagonist in his present situation. We already know he's a man to fear and respect. In this case, scenes that are extraneous to the story's main plot can be eliminated or saved for a possible sequel providing, of course, the first film is commercially successful.

In both the novel and the adaptation of *The Godfather*, Sonny is having an affair with Manicini. But in the novel, we learn what happens to her after Sonny's death. How she starts a new life in Las Vegas, marries a man named Dr. Siegal, and so on. This information is omitted from the screenplay because, after Sonny's death, her character does not add any additional information to the main plot line. In the movie, she only serves to further Sonny's character development. He's a ruthless man who cheats on his wife. After his death, further character development of Sonny is unnecessary.

Maintaining the Pace

When using a short story, scenes are often added to maintain pace. When using a novel, scenes are often omitted to maintain the pace of the screenplay and hold the audience's attention. Again, in *The Godfather* we have the lawyer, Hagen (Robert Duvall), meeting the Hollywood producer, Jack Woltz, to convince him to give Johnny Fontaine a part in his new movie. In the novel, this takes several long meetings spread out over various locations. In the screenplay, the main meeting takes place at the studio and then at Jack's house where we discover his priceless race horse. The other meetings are only alluded to in the dialogue. The screenwriter omits these scenes to move time along. This becomes a minor incident in the screenplay. It reveals Vito's ruthlessness, but scenes for character revelation shouldn't be allowed to slow down the pace of the story. The writer eliminates duplicate scenes, allowing the action of the story to continue.

In the adaptation of *The Graduate,* we see many examples of scenes being omitted to speed the story along. In this story, the protagonist, Ben, is propositioned by Mrs. Robinson, one of his parents' friends. Ben is so shaken up by Mrs. Robinson's initial seduction attempt that he takes a long road trip, hitching into Northern California where he works at odd jobs before returning home. Later, after sleeping with Mrs. Robinson, Ben falls in love with her daughter Elaine. In the screenplay, the writer jumps from the seduction scene in Mrs. Robinson's bedroom, where Ben rejects her, to Ben's change of heart when he invites her for a drink at the Taft Hotel. In the novel, there is also an interesting scene between Ben and his father following Ben's return from his road trip. Talking about Mrs. Robinson, in Charles Webb's novel (1963), Ben's father says: "She's a funny one, Ben. There's something about her that makes anybody feel uncomfortable. She's devious. I don't think she was ever taught the difference between right and wrong" (p. 77).

It's not necessary for the screenwriter to include this scene. Not only would it slow the action down and prevent the connection of the proposal scene to the seduction scene, but it is redundant. Film is a visual medium. When Mrs. Robinson has Ben take her home and bring her coat upstairs, we see how uncomfortable she makes him feel. It's obvious from her actions that she's devious. The novelist tells us this; the screenwriter shows it to us.

Scenes involving Ben and Elaine are also omitted. In the novel, Ben takes Elaine to three clubs before taking her to the strip joint. In the screenplay, we go only to the strip club. Later, when Ben follows Elaine up to Berkeley, the writer omits several scenes involving Ben's search for a place to stay, his move from one hotel to a boarding house, his initial following of Elaine, and his first meeting and conversation with her on campus. Instead, the writer jumps from Ben's arrival at Berkeley to Ben following Elaine on the bus to the zoo. This scene is similar to the scene we find in the book, but it comes later in the screenplay. We also lose scenes involving Ben's move to Canada and Ben's father's discovery of the affair with Mrs. Robinson. The writer eliminates dialogue about Ben flying down to L.A. to get his birth certificate. In the movie, he has it with him.

In *Ordinary People*, the screenwriter eliminates many scenes to prune and move the story along. The incident of Conrad being picked up by the women in the library is omitted because it slows the story down. In the novel, Conrad is a loner whose relationships with women are explored. The screenplay focuses on his problem with his brother's drowning, not his lack of romance. In addition, all the scenes involving backstory about Cal and Beth are also eliminated. Often, scenes that involve extensive philosophizing are omitted.

The scenes involving Cal's philosophy of marriage are cut in Guest's novel (1976): "Two separate distinct personalities, not separate at all, but inextricably bound, soul and body and mind to each other" (p. 149). Beth's views on marriage are also omitted: "I would never come back—not for a house in Glencoe, not for the children, not for anything. It's too humiliating" (p. 160). The screenwriter omits these scenes because he doesn't need to be overt with the thematic content of the story.

Philosophies and overt thematic content don't belong in a screenplay; they slow it down. The protagonist has a problem; it worsens into the crisis at the end of Act Two, and intensifies during the climax in the third act. Anything that stops this progression must be omitted. In *Jurassic Park,* many scenes dealing with Malcolm's philosophy of chaos are cut. They are unnecessary and would slow down the pace of this action–adventure movie. Instead, the screenwriter uses two key scenes to state Malcolm's philosophy and then visually demonstrates it to us in the climax scene. A random act of the Rex killing the other dinosaurs stalking the protagonist saves the day.

CONDENSING SCENES

In addition to omitting scenes, the screenwriter must shorten and condense the existing ones that he thinks are essential to the story. Sometimes this involves shortening specific scenes or even combining two scenes into one. In "Barefoot in the Park," the opening scene between Paul and Corie is shortened. Dialogue about Paul feeling sexy in the middle of a conference, missing Corie, and being on a regular schedule is cut in the screenplay. In *Sophie's Choice*, there is a big fight scene between Sophie and Nathan. In the novel, the action takes place over two scenes. First we see Stingo hearing noises of the two fighting and then, later, when the fighting escalates, Stingo witnesses it outside his room. In the screenplay, the fight takes place at one time and in one place. This saves time but also serves to increase the dramatic intensity of the moment without dissipating it over several scenes.

In *Slaughterhouse Five*, the problem of multiple time periods, with Billy Pilgrim time tripping from one location to another, is solved by combining all the scenes into three time periods: Billy in Dresden during the war as a young man; Billy as a small town optometrist, married, and just starting out; and the Tralfamode period when the middle-aged

Billy is involved with Montana and the baby. Even though the screenwriter jumps from one period to the next and back again, these three periods correspond to the beginning, middle, and end of Billy's life. The story has a beginning, middle, and end—one of the key ingredients of any successful screenplay.

Consider the following scene from the Webb's novel (1963, pp. 54–55), *The Graduate,* when Benjamin meets Mrs. Robinson in the bar prior to their sleeping together:

> They drank quietly, Benjamin smoking cigarettes and looking out the window . . .
> "You've been away," Mrs. Robinson said.
> "What?"
> "Weren't you away for a while?"
> "Oh," Benjamin said. "The trip. I took a trip."
> "Where did you go?" Mrs. Robinson said, taking a sip of her martini.
> "Where did I go?"
> "Yes."
> "Where did I go," Benjamin said. "Oh North. I went north."
> "Was it fun?"
> Benjamin nodded. "It was," he said. "Yes."
> Mrs. Robinson sat quietly a few moments, smiling across the table at him.
> "Darling," she said.
> "Yes?"
> "You don't have to be so nervous, you know."

In the screenplay, all this talk about the trip is cut, with the writer jumping to the next relevant line in the screenplay, "You don't have to be nervous." The following dialogue about Ben's trip and the forest fire up north continues in the novel's scene (Webb, 1963, p. 55):

> "Yes. The big forest fire up there. You might have—you might have read about it in the newspaper."
> She nodded.
> "It was quite exciting," Benjamin said. "It was quite exciting to be right up there in the middle of it. They had some Indians too."
> "Did you put it out?"
> "What?"
> "Did you get the fire out all right?"

This dialogue is omitted in the screenplay, with the writer skipping to the next relevant line on page 55:

> "Did you get us a room?" Mrs. Robinson said.

Compare this dialogue to the scene from the Henry screenplay (1967, pp. 43–44):

INT. VERANDA ROOM—NIGHT

In the door to the lobby in the b.g. is Mrs. Robinson. She pauses, looks into the room, sees Ben and starts toward him. Ben is looking out the window. He does not see her approach.

MRS. ROBINSON
Hello, Benjamin.

BEN
Oh. Hello. Hello.

He rises quickly.

MRS. ROBINSON
May I sit down?

BEN
Of course.

He pulls out a chair, for her . . .

MRS. ROBINSON
May I have a drink?

BEN
A drink? Of course.

Ben looks toward a passing WAITER and raises his hand. The waiter pays no attention. Ben looks back at Mrs. Robinson apologetically.

BEN
He didn't see me.

MRS. ROBINSON
Waiter!

The waiter stops in his tracks.

MRS. ROBINSON
I will have a martini . . .

The waiter moves away.

```
                    MRS. ROBINSON
          You don't have to be so nervous, you know.

                    BEN
          Nervous. Well, I am a bit nervous. I mean it's—
          it's pretty hard to be suave when you're— . . .

He shakes his head.

                    MRS. ROBINSON
          Did you get us a room?
```

The screenplay scene is tighter. It moves quickly, focusing first on Ben's nervousness. This emotion is complicated by the problem of getting the room and the next scene with Ben and the desk clerk. Even the routine with the waiter in this scene is changed from the novel. In the novel's scene, Mrs. Robinson wants a drink, but the waiter doesn't see Ben motioning to him. Benjamin watches and finally gets his attention. In the screenplay, Ben tries to get the waiter but is ignored. Suddenly Mrs. Robinson snaps her fingers and the waiter comes running over to them. The screenwriter has taken advantage of this incident to show us the difference between Ben and the older, more experienced Mrs. Robinson. The fact that it is done visually with a snap of the fingers works perfectly for the film and in one moment gives us the contrast between the two characters we need. The novelist takes three pages of description and narration to achieve the same effect.

COMBINING CHARACTERS

Another method the screenwriter uses to tighten the story involves combining characters. In *Slaughterhouse Five*, for example, the character of Paul Lazarro, the man who was with Billy in Dresden, and ultimately kills him, is a combination of the novel's Paul Lazarro and the soldier, Roland Weary, whose foot Billy steps on. In the film, Paul vows vengeance when Billy steps on his foot. Edgar Derby, the soldier who gets shot for stealing, is a combination of Edgar and O'Hare, Billy's best friend in the novel. Thus, when Edgar gets shot in the movie, the moment carries more dramatic intensity. More is now at stake for Billy when Edgar is shot, because Edgar is not just another soldier in Dresden.

In the short story version of "Shawshank Redemption," there are several wardens who act as the antagonist for Andy. In the movie, all these wardens are combined into one specific antagonist, Greg Stammas. This combination helps unify the story and focus the conflict. It also makes the ending of the movie more dramatic. In the short story,

Andy escapes and uses the money a friend stashed away to make his escape to Mexico. In the movie, Andy visits several banks and takes the warden's illegal funds. He then sends information to the police, which results in the warden's arrest and suicide. The screenwriter has a more interesting ending. The protagonist wins over the antagonist and justice is served. The movie audience leaves satisfied.

FURTHER ELIMINATION AND ADDITION OF SCENES

In most cases, the adapter is forced to eliminate scenes in order to tighten structure. Most of these scenes involve backstory, such as the first three chapters in *Kramer versus Kramer*, in which we learn of Ted's romance with Joanna. Also, in *Midnight Cowboy*, entire sections are cut—the chapters dealing with Joe in the Army, the death of his grandmother, his move to Houston, his job at the Sunshine Cafe, and his mysterious homosexual encounter with the strange character of Perry who comes to visit him each night. In the novel, these adventures serve to deepen and focus the character of Joe, but the screenwriter must find other ways to show these same character traits. In the screenplay, the writer moves from the scene of Joe in his bedroom getting dressed to the scene of him quitting his job at the Sunshine, to his trip on the Greyhound to New York.

What takes 111 pages in the novel is accomplished by page 6 of the screenplay. To show Joe's narcissistic fascination with himself, the screenwriter adds the scene of Joe getting dressed in his hotel room and crosscuts it with the scene from the Sunshine Cafe. The effect gives us a quick insight into Joe by using the visual image while moving the story along to New York where the character's problem intensifies.

All the backstory we get over Joe's earlier years with his mom and then his grandmom and her beaus in New Mexico is given to us in the first scene in Salt's screenplay (1969, p. 1), which runs for less than a quarter of a page:

```
EXT. DRIVE-IN MOVIE SCREEN—DAY

A blank whiter frame-angle widening to show an empty drive-
in movie screen—towering over a lonely landscape—a solitary
flicker of movement—TINY JOE BUCK swinging alone in the
playyard below the screen . . .

an old cowboy voice singsonging O.S.—a buckaroo's dream of
heaven—his experiences among the angels continuing through
TITLES superimpose over . . . (p. 1)
```

This is, of course, the character of Woodsy Niles from the novel, one of the grandmother's beaus, the closest man to a father for the born out-of-wedlock Joe Buck.

Backstory on Mitch in *The Firm* has been carefully cut down. In the novel, we learn about his mother's background, her life in a trailer park in Panama City, her check from a mine explosion, the death of her eldest son in Vietnam, and, of course, the imprisonment of her other son. In the movie, the only relevant backstory is about the imprisonment of her son.

Sometimes scenes are added in screenplays when the writer uses narration or description to reveal a character action. In "The Rocking-Horse Winner," Lawrence describes the mother's obsession with money and how this obsession began to haunt the house and her children. In the film, the writer adds scenes to show us the children being haunted by the mother's need for more money. Many fiction writers tell us that the character is feeling something; the screenwriter dramatizes it. In *Wuthering Heights*, Emily Brontë goes to great lengths to describe Cathy's emotions when Heathcliff leaves. In the screenplay, Cathy runs out into a raging storm. The wind and rain symbolize her conflict between her love for Heathcliff and her need for the respectability she hopes to gain by marrying Edgar.

Other additions to screenplays are used to fill in scenes between subplots that have been omitted, or to fill in and connect the main plot line when characters have been omitted, or to expand a story line. The opening of the short story "The Sentinel" has been expanded into the full-length screenplay *2001*, with scenes that were not in the original story. On the other hand, sometimes scenes are added, such as in *Midnight Cowboy*, to explain a character's motivation or, more likely, to achieve a dramatic effect that is only described in the novel. In *Wuthering Heights*, the scene of Heathcliff smashing his hand through the window of the stable when he learns of Cathy's marriage to Edgar is added for dramatic purposes. It does not appear in the novel. **[Exercise 4]**

BEGINNINGS AND ENDINGS

As mentioned earlier, often, the screenwriter must find a new starting point for her screenplay, eliminating several chapters or pages of the source material to find a troublesome situation with which to begin. The screenwriter takes an incident that occurs two or three chapters into the book as her starting point, such as in *Ordinary People* or *Kramer versus Kramer*. Sometimes a writer takes an incident that occurs later in the book and places it at the beginning to set up the initial problem. Such is the case in the adaptation of *Dune*. In the novel, Paul becomes a threat to the Empire midway through the book. In the screenplay, the writer changes this structure, making Paul a threat at the very beginning and then adds obstacles for Paul to make his problem worse. Again, the consideration of rising action and holding the audience's attention comes into play.

Openings are sometimes changed to make the story more visual. In *The Silence of the Lambs,* the novel opens with Agent Starling on her way to Crawford's office to learn of her impending assignment. The first draft of the screenplay followed this structure but by the final shooting script, the opening had changed. The movie opens instead with Clarice out on the training range. The novel talks about the grass in her hair and the stains on her windbreaker but the movie shows us how she got them by running through the obstacle course. It's a more interesting opening and definitely more visually pleasing to watch. But, more important, it tells us things about the protagonist. We see she's still a trainee and we see her in action. We see she is dedicated and full of perseverance—necessary traits for the impending battle with Buffalo Bill.

Endings are also important in a screenplay. Often endings are changed to make the scene more visual; for example, in *Kramer versus Kramer*, a telephone conversation in the book is substituted for a face-to-face encounter. Or, in *Dune*, the discussion about Paul and his wife is substituted with the more dramatic scene of Paul as the superbeing of legend.

Sometimes endings are changed to add a unity to the screenplay. This is the case in *Treasure of the Sierra Madre*. In the novel, Dobbs is killed by three Mestizos Indians. In the screenplay, Dobbs is killed, ironically, by three of the bandits who had jumped him earlier at the mine. As Houston himself says in Belmer (1986): "The primary purpose was to give the moving picture a certain dramatic unity which novels don't necessarily require. . . . Pictures and plays seem to fall apart when new characters and incidents are introduced as final scenes" (p. 296).

Endings are occasionally changed to maintain the sympathy of the protagonist. In *The Firm*, for example, the novel ends with Mitch stealing the money from the FBI and escaping an intense hunt by the mob in Florida. Mitch escapes to the Bahamas where he lives comfortably off of the stolen money. In the movie, the ending is changed to make it more appealing to the audience. Mitch outwits both the FBI and the mob and is able to maintain his career as a lawyer. He thinks of a clever scheme for having the members of the firm indicted for false billing rather than violating the attorney–client privilege. Mitch is not disbarred and proves to be more clever than the FBI or the mob without doing anything illegal. The audience, who has been rooting for him to succeed, is not cheated at the end of the movie.

In an earlier scene involving Mitch's infidelity with a woman on the beach, changes have also been made to maintain the character's sympathy. On page 159 of the novel we find: "She unsnapped something and removed her skirt, leaving nothing but a string around her waist and a string running between her legs." Mitch succumbs to simple lust. In the movie, he resists one attempt to seduce him and only becomes involved after saving the woman from an attack on the beach. This incident is added to make Mitch's affair more acceptable. **[Exercises 5–6]**

SUMMARY

The writer who understands structure and does not get bogged down among the many subplots and characters in the novel is most likely to produce the most commercially successful screen adaptations. As with characterization, faithful adaptation of the structure of the novel is an ideal that often cannot be realized. The screenplay writer has to make the structure his own even if that means cutting and eliminating scenes or adding new ones. The next chapters show how social issues and dialogue are important elements in this process.

EXERCISES

1. Paddy Chayefsky is best remembered as the writer of *Network* and *Altered States*, but in an earlier work, Chayefsky's 30-minute teleplay *Printer's Measure*, we explore the relationship between a young boy who just graduated from a technical high school and an old man who prints by hand. The man is an artist and the boy wants to emulate his skills. The old man agrees to teach him the trade, but the issue of technology enters the story. The time is the early 1900s and printing is just starting to be done by machine. The boy ultimately chooses the machine over the handprinting, artistic method of the old man. What starting point would you pick for this story? Be specific, set the scene, and write the opening dialogue. Remember you only have a half-hour to tell your story. Check the appendix to see how Chayefsky dealt with this problem.
2. Another Chayefsky teleplay, *Holiday Song*, was adapted from a magazine article he read dealing with a photographer who, while riding a Brooklyn subway, meets, at separate times, a man and woman, both concentration camp victims, who lost track of their spouses during the war. The photographer concludes that they are husband and wife and reunites them. Chayefsky changed this story by making his main character not a photographer, but someone who has lost his faith in God.

When he mistakenly gets on the wrong subway two times in a row, therefore meeting the separated husband and wife by accident, he concludes that God had been with him, and, in fact, guided him to that specific subway. His faith in God is restored and he experiences character growth. What starting point would you pick for this story? Who would your protagonist be and how would you dramatize his problem (loss of faith) in the first scene? Write your scene and discuss the reasons for your starting point in class.

3. Watch several movies and determine why the writer picked a particular starting point for the story. What does the first or second scene of the movie reveal about the protagonist's problem, goal, or conflict?

4. Take the following climactic scene from Ann Beattie's *Chilly Scenes of Winter* and condense it. In this story, Charles has been in love with Laura who has finally left her husband. He goes over to her house to find out if she's coming back to him. In the novel, the scene runs 11 pages—289–300. This is just a small excerpt, mainly the dialogue:

> "Come in . . ."
> "Coming in, I started remembering that dessert you used to make with the chocolate and the oranges . . ."
> "Oh I know the one you mean. You can come over sometime and I'll make it for you."
> "Sometime?" What is she talking about?
> "Tomorrow?" he says.
> "Tomorrow? I guess so. If you want to," she says . . .
> "You've got a roommate?"
> "Yes she's at the library. She's in graduate school."
> "Oh. Well, what are you doing?"
> "Looking frantically for work."
> "Why don't you come back to the library?"
> "I don't want to." . . .
> "Are you looking for another job like that?"
> "I wouldn't care. I've just got to get a job."
> "Sam's out of work. You remember Sam?"
> "Of course I do." . . .
> "That's too bad." . . .
> "I love you. I want to know what's going on."
> "I can't tell you. I don't know myself. . . ."
> "You can stay with me."
> "No. I just want to think things out for a while. . . ."
> "Don't you remember how we ate dinner together and went to the movies and . . .?"
> "I remember it perfectly."
> "Then move in with me." . . .
> Laura shrugs.
> "Somebody else?"

> "Nobody I'm serious about."
> "Who, Laura?" . . .
> "A taxi driver." . . .
> "Where did you meet a taxi driver?" . . .
> "In his taxi." . . .
> "He picked you up?"
> "I don't want to get into an argument. I'm feeling very low. . . ."
> "You get in a goddamn taxi and you let the driver pick you up?"
> "So what?"

Condense and finish the scene. Check the appendix to see what the screenwriter did with this long scene.

5. Check the endings of various adapted novels—*Ordinary People, The Firm, Jurassic Park, Midnight Cowboy, The Kiss of the Spider Woman, Dune,* and then rent the movies or read the endings of the screenplays. Have any of the endings been changed? Can you figure out why? Discuss these differences in class.

6. In *Terms of Endearment,* McMurtry's novel ends with the characters gathered at Aurora's house following Emma's funeral. The emphasis is on a conversation between Patsy and Aurora. They are talking about Melba, Emma's friend:

> "I don't know what we're to do with that poor woman," Aurora said. "I'll make Joe talk to her," Patsy said.

The conversation continues regarding Aurora's guilt over her treatment of Emma (McMurtry, 1975, pp. 370–71):

> "She often made me feel I was faintly ridiculous," Aurora said. "Somehow she just had that effect. Perhaps that was why I remained so unremittingly critical of her."

In the screenplay, the scene is changed. Patsy is talking to Flap, and the astronaut (Jack Nicholson) is talking to Tommy, the elder of Emma's boys. Aurora is talking to her maid. Why did the writer make these changes?

5

Kramer versus Kramer

The adaptation of Avery Corman's novel (1977), *Kramer versus Kramer,* into the screenplay of the same name shows all the classic problems—starting point, elimination of characters, change in characterization, character growth, rounding out of characters, different social issue, elimination of scenes, tightening structure, and addition of obstacles for the protagonist—the screenwriter confronts during the adapting process. This chapter examines this adaptation by focusing on the following nine key points:

1. *Find the starting point.* How does the screenplay start differently from the novel?
2. *Change the characterization.* How are Ted and Joanna changed from the novel?
3. *Round out characters.* How is the antagonist made more sympathetic?
4. *Eliminate characters.* How are minor characters eliminated or combined?
5. *Rework the social issue.* Does a change on focus in the social issue change the focus of the screenplay?
6. *Rework the structure.* How does the screenwriter achieve a tighter structure?
7. *Rework the subplot.* How is the subplot changed and tightened?
8. *Eliminate scenes.* Which scenes are cut from the novel and why?
9. *Add scenes.* Which scenes are added that aren't in the novel and why?

STARTING POINT

The screenplay starts at page 44 of the novel. The novel's first two chapters deal with backstory about Ted and Joanna—their preparation for Billy's birth, their experiences on Fire Island and their early dating history, and Ted's friendship with Larry. This elimination provides several useful changes for the screenplay. In the novel, for example, the first chapter is devoted to preparation for Billy's birth. We learn that Ted is a good father: how he wants to participate in the birth process with Joanna; how he wants to do his share with Joanna; how scared he is that he might not be the ideal, good father; how Ted goes to childbirth classes with Joanna and helps her practice her breathing. In Corman's (1977) words, he "was a model pre-daddy" (p. 11). After Billy's birth, Ted did his share as well. He did not stay late at work or flirt with the secretaries, but instead came home each night. At home, he also helped out. He took Billy out of the house to the park, changed his diapers, and washed him.

The Ted of the novel is a far different Ted than the one in the screenplay. Here, Benton's (1978) Ted is a basic workaholic: a Ted who spends more time at work or drinking with the guys than he does at home helping Joanna; a Ted who, in the screenplay's third scene, comes home late from work and is sorry he didn't call—"I was busy making a living" (p. 6); a Ted who is a totally inadequate father in Act One after Joanna leaves him; a Ted who has trouble making his son breakfast the first morning Joanna is gone; a Ted who cooks his French toast with egg shells in it, burns the toast, and winds up screaming at his son.

The reason for this important change in characterization is twofold. One relates to the starting point of the screenplay. The screenplay must begin with a troublesome situation, that is, Joanna leaving Ted and making Ted care for their son. The novel begins leisurely with Ted practicing to be a good father. But the screenwriter chose to focus on the issue of parenting, that is, who is the better parent—the mother or father? To answer this question, the writer picks a later starting point in the novel, which also presents a problem for the protagonist.

Second, this change provides the protagonist with an opportunity to experience character growth. At the beginning of the screenplay, Ted is a very inadequate daddy, not the "model daddy" of the novel. As he learns what is important to him during the course of the screenplay, namely his son and how much he loves him, Ted changes and grows. As he becomes a more capable, loving father, we become more involved with him. He gains our sympathy and we root for him in his custody suit with Joanna. When he wins at the end, we share his triumph because he has changed into the caring, loving Ted whose only concern is his son, Billy. He's a great guy and we feel that we are like Ted—great caring people.

As his friend, Thelma, describes him in the courtroom scene, "He reads to Billy. They play together. They talk all the time. He's a very devoted father" (Corman, 1977, p. 114). He's so devoted that even Joanna experiences character growth in the screenplay. She goes from a woman who walks out on her husband, to someone who admires Ted for what he has accomplished with their son. She admires Ted's parenting so much that she is willing to experience the pain of allowing Billy to stay with Ted. Not only do we admire Ted at the end of the movie, but we are crying for Joanna as well.

CHANGE IN CHARACTERIZATION

There is a change in Joanna's characterization as well. In Corman's novel (1977), Joanna's motivation for leaving Ted is the boredom of raising a child. As she tells her friend, "I love my baby, but basically it's boring" (p. 24). While Ted is off at the office experiencing the thrills of working in New York, "Joanna was at home, trying to stay engaged while Billy was building blocks into a garage" (p. 26).

In the screenplay, Joanna is different from the tennis fanatic of the novel. Her reason for leaving is Ted's neglect. Ted is the initial heavy in the screenplay. He works all the time, is late coming home, and never helps out with Billy. Joanna is the wronged figure, the person Ted never communicated with, the person driven to the desperate measure of leaving her child in order to regain her sanity. Thus, the screenwriter accomplishes one goal by changing character motivation. He makes Joanna more sympathetic at the beginning of the screenplay.

In the novel, Joanna isn't very likable for leaving her son. Also, by making Ted less sympathetic, Joanna gains more sympathy from the audience. Now this is the essence of real drama—the audience is caught between two characters who, in a sense, are both protagonists in this story. There is no clear "good guy" or "bad guy." Who should we root for? Sure, in the beginning Ted is more of the heavy, but this quickly changes as he tries to cope with his situation. Joanna then becomes the heavy when she writes Billy her good-bye letter, but this also changes when she comes back into the story in Act Two. Now she has her life back together and is desperate to see her son again. The audience remains hooked into the third act trial scene. Who should really have custody of Billy? Both parents are now sympathetic and both have good reasons for keeping Billy, as elaborated in the social issue dialogue dealing with the rights of the father versus the rights of the mother. The audience is still hooked. No matter who wins or loses, the audience is moved because the screenwriter has made it possible for us to like both of his main characters.

Another effective change in characterization involves O'Connor, Ted's boss. In the novel, O'Connor has a good working relationship with Ted. Ted loses his job because the company they work for has gone out of business. In fact, it is O'Connor who gets Ted another job with a new company. "O'Connor could not promise anything—he was just getting established, but he wanted Ted to work for him" (p. 148).

In Benton's screenplay (1978), O'Connor provides more of a problem for Ted and a stimulus to building conflict between Ted's obligations as a parent and his commitment to his job. O'Connor suggests that Ted give his kid to his parents to take care of. As he warns Ted, "If I can't count on you a hundred and ten percent, twenty-four hours a day, seven days a week because you're worried about a kid with a running nose, there may be a problem" (Benton, 1978, p. 24). Already, by changing O'Connor's attitude, the screenwriter creates more tension and conflict between the characters. This tension makes the scene more interesting.

At first, Ted continues to assure O'Connor that he will be there for him 110%. But as the story progresses and Ted becomes more involved in caring for his son, we begin to see small changes, especially in Ted's appearance: "His hair is longer and less neatly cut. There is a grape juice stain on his jacket. He is carrying . . . a shopping bag and some complicated construction that Billy had made for school" (p. 42).

Increasingly, Ted becomes less able to balance these two parts of his life, causing increased tension and conflict between him and O'Connor. As O'Connor angrily says during a business meeting: "If it's alright with Mother Kramer . . . can we get down to work now?" (p. 42). Ultimately, O'Connor fires Ted as Ted's attention to Billy increases. The problem is that Ted needs his job to ward off Joanna's impending custody suit. Without a job he has no chance of retaining custody. Another obstacle is put in the protagonist's path.

ROUNDING OUT THE CHARACTER

Interestingly enough, structurally, this minor change in O'Connor's attitude serves to increase the tension and complications that Ted has been having ever since Joanna left him. The writer tends to make O'Connor somewhat justified in firing Ted. This is made apparent to us during a business lunch scene, when Ted, forgetting whom he's with, cuts his client's meat into little pieces. The fact that O'Connor has some justification for firing Ted makes his character well rounded and not just a classic "bad guy" antagonist.

This dimensional representation even adds some credence to O'Connor's position that Ted should concentrate on his work and leave the raising of his child to his parents. The issue of whether Ted should work or be a parent is a valid one for the parameters of this story. Who is right? In the well-constructed story, all positions have some validity. In good drama, there is no clear-cut, black-and-white answer.

ELIMINATION OF CHARACTERS

Another device the screenwriter uses to tighten the structure is elimination of the many superficial characters in the story. In the novel *Kramer versus Kramer*, much time is devoted to Larry, Ted's friend from Fire Island. But the screenplay is not concerned with the singles' scene and Larry's technique for picking up women. Instead, it focuses on the issue of parenting. Larry is unnecessary. We also lose Joanna's concerned parents, Ted's parents, his older brother Ralph, and Thelma's husband, Charlie, who is just alluded to in the script.

The screenwriter moves time along in the subplot by having Thelma divorce Charlie. The novel deals with his adultery and subsequent divorce. The elimination of all these superficial characters, again, tightens the story, allowing us to focus on the main plot line of Ted trying to cope with raising Billy.

Even Thelma, a rather old-fashioned name for a character, becomes the more modern Margaret in the screenplay. Note that many times a character's name is changed from the novel to the screenplay. The character of Joanna's boyfriend who tries to intercede in the custody battle is also dropped as the writer moves quickly from Joanna's return to the custody trial—the climax of the third act.

As a divorced man, all Ted's dates are eliminated in favor of keeping one character, Phyllis, who becomes Ted's only romantic involvement. Significantly, she works where Ted does. This serves to unify the story and make the point concerning Ted's social life with one character rather than several. Again, the emphasis is not really on Ted as a single man, but on Ted as a father. The issue of dating is only briefly explored, and then only later in the custody scene in connection with the social issue—Is Ted a fit father if he slept with Phyllis while raising Billy?

SOCIAL ISSUE

The foundation and backbone for most screenplays is the social issue. In the novel, the issue is really twofold. What problems do the recently divorced or single person face in today's fast-paced dating world and who is the best parent, a mother or a father? These questions get equal play in the book.

The beginning of Corman's novel explores the entire question of dating when it gives us backstory about the initial relationship of Ted and Joanna. Early in the novel, the writer (1977, p. 18) describes Ted as follows:

> He was reasonably unused to the singles' predicament they both shared not as detached as Vince, an art director, who had been standing around her desk and who told her he was bisexual, and not as desperate as Bob, a media supervisor who was on the verge of divorce.

Later, when he meets Joanna, they both confide to one another how sick they are of dating and the social scene. When Joanna leaves him, Ted is forced to confront the social scene again: "His wife had left him and if your wife leaves you, somewhere along the line you have to deal with single women" (Corman, 1977, p. 18).

He was back in the dating world of one-night stands and singles' bars. In the novel, an entire chapter is devoted to being single when Ted takes Billy back to Fire Island for a vacation and meets a woman on the beach. We learn how out of practice Ted is, and how difficult it is to have sex when you have a child in the same room with you. We also learn about all the walking-wounded singles in this chapter—the woman who had a nervous breakdown because she couldn't find anyone to date and couldn't get out of her chair one night, the psychiatrist who never stops reading, and, of course, Larry who uses his station wagon to pick up women.

The novel also devotes time to Ted's relationship with a 20-year-old and the question of her being too young for him. Later, after his divorce, Ted winds up cruising the singles' bars with Charlie and ends up, at one point, in a macho leather bar. In the novel, Ted's reaction to the new singles' scene becomes so intense that we find him visiting a shrink and discussing whether he should be dating.

In the screenplay, the writer eliminates the focus on the singles' scene, choosing to focus, instead, on the problems of single parents. Again, the issue is whether Ted is capable of being a single parent and having a successful career. As Benton's O'Connor says, "I can't let your family life interfere with work" (1978, p. 51). Of course in the middle of this conversation with Ted reassuring O'Connor, Ted gets a call from Billy about how much TV he can watch per night.

In the screenplay, the real crux of the question is who is the better parent, Joanna or Ted? In the novel, the custody scene is longer because there are more witnesses. We hear from O'Connor, still on friendly terms with Ted in the book. We hear from Ted's housekeeper, Joanna's father, Charlie, Ellen, Billy's elementary school teacher, and Ted's sister-in-law, Sandy. In the screenplay, the testimony is confined to Margaret, Joanna, and Ted. The writer condenses the scene by focusing on three key people.

Basically, the dialogue is similar in both novel and screenplay except for Ted's testimony. In the novel, Corman uses narration. "What passed in the courtroom was a description of nothing less than a man's life" (1977, p. 227). In the screenplay, this description becomes a very emotionally lengthy piece of dialogue giving Ted's position on the social issue of parenting. Note that lengthy dialogue should normally be avoided except toward the end of a screenplay, when the protagonist usually reveals her or his feelings in a long emotional speech. In Benton's screenplay (1978), Ted says, "I guess I've had to think a lot about whatever it is that makes somebody a good parent: constancy, patience, understanding . . . love. Where is it written that a man has any less of those qualities than a woman?" (p. 115).

In Corman's novel (1977), the summary of Ted's life raising Billy and his fitness as a father is boiled down into the summary given by Ted's lawyer, "'Give the man his kid'—he seemed to be urging the judge" (p. 226). The argument for and against Ted as a father is more developed in the screenplay because the issue of parenting is more focused and presented from different points of view. It's a complex question and an emotional one. The fact that each point of view is developed and valid adds to the improved quality of the screenplay over the novel.

STRUCTURE

A screenwriter must condense the long and often rambling structure of the novel or expand the short, compacted short story. In most cases, a writer has to condense a novel's 400 to 500 pages down to 120 pages of script. For *Kramer versus Kramer*, 247 pages are condensed into 132 pages of script.

The first step in condensing is finding the proper starting point. As previously mentioned, the scriptwriter chose to eliminate the first three chapters of the novel and begin in the middle of Chapter 4. The criterion for his decision is simple. The scriptwriter needs to begin with a troublesome situation that can escalate into the problem of the story. In this case, the troublesome situation is that Joanna is leaving Ted on page 6. It's only a troublesome situation in the script because Ted is convinced that Joanna will return and his coping with her leaving is only temporary. This, however, becomes a full-blown problem on page 26 of the screenplay when Ted receives Joanna's letter to Billy in which she makes it official that she will not be returning. This speech is similar to the one we find in the novel (Corman, 1977): "I have gone away because I must find some interesting things to do for myself in the world. . . . I will always be your mommy and I will always love you" (p. 54). This marks the end of Act One. Ted has a definite problem. Joanna will not be returning and Ted must face the responsibility of taking care of Billy permanently.

Act Two begins with the protagonist temporarily coping with his situation. In the screenplay, we see Ted talking to Margaret to find out Billy's pediatrician. In the next scene we see him shopping with Billy and learning the correct products to buy. But since the problem must once again emerge in Act Two, if our story is to continue, we find that by page 30 Ted is having problems balancing his job and caring for Billy. In Benton's living room scene, Ted is working when Billy suddenly spills some juice on Ted's important papers:

> Jesus Christ! I've got a major, I mean
> major presentation coming up next week.
> There's a lot riding on this, pal. (1978, p. 30)

In the novel, the pace is much more relaxed; Joanna doesn't leave Ted until page 44. The novelist has three full chapters to devote to character development and backstory about Joanna and Ted, and even to develop the book's social issue. On page 54, we get Joanna's letter. The letters are basically the same, but the screenwriter, whenever possible, omits extraneous material that does not enhance the point he is trying to make. In this case, the excerpts about Joanna sending Billy toys and birthday cards and sending him kisses when he's asleep, and the comparison between Ted and Billy's teddy bear are omitted.

The writer remains consistent to this omission by later omitting the scenes involving Joanna's return to New York with presents for Billy and his excitement at seeing her again. Instead, in the screenplay, Joanna only sees Billy when she returns to New York to file for custody. Again, the screenwriter moves the story along, eliminates duplicate scenes, and condenses information that is not essential to the emotional feeling he's trying to convey. He doesn't have time to be indulgent in 120 pages.

SUBPLOT

The key to successful structure in the *Kramer versus Kramer* screenplay lies in economy. One way to simplify any novel is to simplify the novel's many subplots. In the novel, we have the main subplot involving Thelma and Charlie, and we have a subplot involving Ted and his parents and brother Ralph, his trip to Florida, and dinner and baseball game with Ralph. We also have a subplot involving Ted and his relationship with Joanna's parents and their frequent trips to New York to see Billy and check up on Ted. We also have Ted's relationship to the 20-year-old and his single adventures on Fire Island with Larry.

In the screenplay, the writer eliminates all these extraneous characters and subplots, choosing to focus on the Thelma–Charlie subplot. Again, in the interest of speeding the story along, the writer changes the structure of the subplot by condensing time. In the book, we see Charlie and Thelma's relationship prior to Joanna leaving and during the early days of the breakup, when we learn that Charlie has been cheating on Thelma. The screenwriter, on the other hand, begins his subplot with Thelma—Margaret—divorced from Charlie and coping with the raising of her own child. Immediately this puts the subplot on a more parallel course with the main plot. Both Ted and Margaret are single parents coping with raising their children. How Margaret successfully copes, compared to Ted's inadequate efforts, ties more closely into the social issue. After all, Margaret is a woman and maybe women should be the preferred parent when it comes to childrearing.

In addition to playing more into the issue of the movie, changing the starting point of the subplot shifts the emphasis from marriage and divorce to parenting. In the novel, Joanna's leaving definitely affects Thelma's relationship with Charlie; it's the beginning of the unraveling of her own marriage. In the screenplay, the question of marriage breakup is not really explored. Since Margaret is already divorced, we don't have to question whether Joanna is the cause of Margaret's divorce. We also lose the single issue once again. Thelma questions the whole singles' scene where one relationship after another is never sustained. In the screenplay, she is past the newness of being divorced and having to date again. The emphasis in the novel is on the single life; in the film, it's on the definition of a family. As Margaret tells us, when a person has children with a man, there's a kind of permanent connection that lasts until the end of your life. If you're married, you're married for life.

ELIMINATION OF EXTRANEOUS SCENES

Apart from elimination of subplots, the screenwriter also eliminates other extraneous novel scenes to tighten the structure. The writer's goal is always to move the story along. Even the relationship with Phyllis is made tighter in the screenplay by having her work in the same office as Ted, eliminating the need for Ted to meet her and build up to dating her, as in the novel. The job-firing sequence is speeded up. In the novel, we get hints that Ted is going to lose his job. In the screenplay, one scene is devoted to O'Connor suddenly firing him. The job search is also tightened. In the novel, we see Ted interviewing at several agencies, spending time in the New York Public Library, and making calls and going on different job interviews. In the screenplay, he goes to one employment agency and one job interview before getting his job. In the book, he gets a second job and then loses that one before he gets his final job.

Other examples of tighter structure include the custody case. In the book, Joanna confronts Ted about custody of Billy at a restaurant. This is followed by a scene with Joanna's boyfriend discussing the same matter with Ted in front of another restaurant. In the screenplay, this scene is omitted and we move to the trial more quickly. By focusing on the more important moments in the novel, the screenwriter is able to maintain the pace, suspense, and rising action so crucial to the screenplay.

Even the ending of the novel is changed. In the novel, Joanna calls up Ted to tell him she has decided that he can keep Billy. In the screenplay, the writer has her give this information in person. It's more dramatic this way. We can see Joanna's pain and Ted's elation. We are torn between the two characters, both likable and in pain.

ADDING SCENES

Sometimes, structurally, the writer will add scenes instead of cutting them. Often, this is done to fill in key moments between important scenes, while keeping the action rising. Sometimes it adds to characterization or develops another point of view. In the screenplay, the only added scene of any importance is the one involving the psychiatrist's visit to Ted's house. In the novel, Ted is visiting the psychiatrist. In the screenplay, the psychiatrist is visiting Billy. This is an important scene because it's the only one that gives us Billy's point of view on the issue of the movie—that is, how good a parent is Ted, really? In the novel, we never know how Billy feels.

SUMMARY

The success of the screenplay adaptation of *Kramer versus Kramer* is apparent for several reasons. By following a tighter structure and eliminating extraneous characters and subplots, the screenwriter is able to keep up the pace of the story and hold the audience's interest. By changing his characters, the writer is able to intensify the conflict of the piece and manipulate the audience between two likable characters, Joanna and Ted. The one-sided characterization, "good guy" and "bad guy," has been eliminated. This change in characterization reinforces the question of the social issue. If both parents are sympathetic and likable, and the case for both sides is so well presented, then who would be the more appropriate parent to raise Billy?

It is this key question that keeps members of the audience on the edge of their seats, waiting for the resolution at the end of the movie. That there is no clear-cut answer is what makes this issue so interesting and this movie so appealing. It's no surprise that this was a box office hit. The question is: Why aren't other adaptations of novels as commercially successful as *Kramer versus Kramer*?

EXERCISES

1. Take the following character description of Joanna from page 4 of Corman's novel (1977) and rewrite it for inclusion in a movie script. Compare your version to the actual screenplay version in the appendix.

> Joanna Kramer was nearly professional in her looks, too slight at five-three to be taken for a model, possibly an actress, a striking, slender woman with long, black hair, a thin elegant nose, large brown eyes and somewhat chesty for her frame.

2. Finish the following dialogue between Ted and Joanna when she confronts him regarding taking custody of Billy.

```
                    JOANNA
Ted . . . the reason I wanted to see you
. . . I want Billy back.

                    TED
You want what?!
```

Check your scene with pages 75–76 of Benton's movie script and this book's appendix.

3. Reread the first three chapters in the novel *Kramer versus Kramer*. Decide what backstory you would include in your adapted version. Using that backstory, write a scene for submission to your instructor.
4. Reread the *Kramer versus Kramer* screenplay, or watch the movie. Can you see moments of foreshadowing in the dialogue? Can you see moments when the dialogue is connected from one line to the next?
5. For the famous trial scene in the movie, add a speech by Billy testifying about which parent he wants to stay with and why.
6. Rewrite the ending of the movie, giving Joanna physical custody of Billy.

6

Social Issue

Abortion, child abuse, alcoholism, drugs, and wife beating are all popular social issues. A screenplay benefits by using such issues to create a strong foundation for its story. In the adapting process, the writer may refocus the social issue. There may be several issues in a novel that need to be narrowed down to fit the tight time constraints of the screenplay. Or the screenwriter might want to deal with an issue that is more contemporary than the one in the novel, one more likely to appeal to a mass audience. Sometimes a screenwriter has a problem with censorship. There is more leeway in what gets printed than there is in what gets shown on the screen. Here, commercialism plays its part in the screenwriter's decision. An X-rated movie is more likely to have a smaller audience than an R- or PG-rated film.

To understand the adapting process, one must first understand the nature of the social issue. How does the issue fit into the structure of the story and how do we show multiple sides of the issue to generate conflict? How do we catch the audience's attention with the issue's various points of view? Finally, understanding and using a social issue successfully is a skill a writer can use whether she is writing an original screenplay or adapting it from another medium.

Successful dramatization of an issue depends on anticipating which issues will be relevant in the future. In *The China Syndrome*, the writer successfully anticipated the crisis in the nuclear energy industry before the problem at Three Mile Island occurred. Picking an issue that hasn't quite reached the level of public awareness makes your screenplay a more interesting commercial property.

Successful dramatization of an issue also depends on picking complex issues that don't have easy answers. The more complex your issue, the more conflict it will generate. Issues that have a high degree of emotional residue will more involve your audience because they root for the character who represents the side of the issue in which they believe.

Successful dramatization of an issue depends on making each side believable. In *The China Syndrome*, one group of characters favored the peaceful use of nuclear energy. Their position was that nuclear energy was safer, less costly, and polluted the environment less than fossil fuels. Other characters in the story believed that nuclear power plants were unsafe and posed a great threat to the general population. In the end, the writer makes his own position apparent when we learn that the plant is indeed unsafe because of poor workmanship during construction. Who is right in this story? In a sense, all the characters are right. In a story with a complex issue, the audience is still thinking about the relevant points even after leaving the theater.

Successful dramatization of an issue creates interesting secondary issues. In the novel or longer short story, the writer often deals with more than one issue. In the adaptation, the writer focuses on one main issue and uses secondary issues to enhance the characters' arguments about the main issue. In *The China Syndrome*, the issue is the safety of nuclear power plants; however, secondary issues, such as poor workmanship, human error, and human greed, are also touched on. These issues add complication to the primary issue.

Nothing is totally black or white. Without the human factor, perhaps these plants would be as safe as some of the characters contend. On the other hand, maybe the human factor is something that needs to be part of any discussion about nuclear energy. All the back-up systems in the world are useless if the equipment hasn't been built properly.

CHANGING THE FOCUS

This chapter looks more closely at how the writer uses a social issue as the foundation for a screenplay and how the issue in the case of adaptation differs from the issue in the novel. Specifically, we look at five different movies; three of them—*Kramer versus Kramer*, *The Paper Chase*, and *The Bridges of Madison County*—are adaptations; *Harold and Maude* and *Roe versus Wade* are originals. The first four movies are theatrical films and the last one is a movie made for TV.

Kramer versus Kramer

In the adaptation from novel to screenplay, the screenwriter sometimes simplifies the social issue of the story. In *Kramer versus Kramer*, the writer omits the problems of being a single man—singles' bars, dating, and one-night stands. Instead, he focuses on the main issue of the novel—the question of who is really fit to be the best parent to a young child, the mother or the father.

The writer prepares us for this question by making the case for both parents a good one. If the father had been a child abuser or the mother an alcoholic, then the answer would be easy. But if both parents are good parents and essential to the child's welfare, then the social issue becomes more complex and thus more interesting.

The fact that our society is characterized by a rising divorce rate, increasing numbers of single parents, and women and men taking on new social roles makes this question not only interesting, but also crucial for all members of the viewing audience. In the 1970s, perhaps, this issue was more relevant than it is today. Judges normally awarded custody of children to their mothers. In the 1990s, many divorced couples opt for joint custody, with men seen more in parental roles such as in the movie *Mr. Mom*. The successful adaptation of *Kramer versus Kramer* in the late 1990s might mean further changes in the issue to bring it up to date and more relevant for today's audiences. The focus could be on the problems of joint custody.

Successful dramatization of any issue, however, means the writer must give us all sides of the issue. In Ted's case, we have the first and second acts of our story, which deal with Ted's skills as a parent. We see him struggle from the inept father of the first scenes (yelling at Billy for not eating his dinner) to the warm, caring, and concerned father (dramatized in the scene of Ted running through the streets to the hospital with Billy who is bleeding from his fall on the playground).

We see Billy's point of view, his dependency on his father, and fear that his father will leave him. The cute scene of father and son eating donuts for breakfast and both reading at the kitchen table shows how close they have become. As Benton's (1978) Margaret says in a courtroom scene, "He reads to Billy. They play together. They talk all the time (*tears start*). He is a very . . . kind man . . . a very devoted father" (p. 119). Margaret, who had been on Joanna's side when Joanna left Ted, is now supporting Ted. She, like the viewers, has seen what a good father Ted has become and why the judge should not take Billy away from him.

But what about Joanna? If Ted is such a good father, where's the conflict? Why should Joanna get custody of Billy? Shaunessey, Ted's lawyer, tells us why when Ted comes to him for advice. The courts usually decide that the best place for a child is with his or her mother, even if the mother has given up her custodial rights. This is the legal position and is borne out by the court's decision to give Billy to Joanna at the end of the story. Yet these are modern times. Fathers have rights too. The writer, therefore, has to build up Joanna's case so that emotionally she has as much right to

Billy as Ted. We have to see why both Ted and Joanna should have custody of Billy. We have to go beyond the legal decision of the courts.

The screenwriter maintains the other side of his social issue by keeping our sympathy for Joanna even when she's not on the screen. He starts off by making Ted the cause of the breakup in the movie and by having Margaret support Joanna. We see, on the screen, how painful it is for Joanna to leave Billy and how much she loves him. We see how Billy misses his mother because he sleeps with a picture of her by his bed. We like Ted, but we know that the boy also needs his mother. Later, Ted admits that the breakup of the marriage is his fault. One characteristic of this issue is that children often think it's their fault when a marriage breaks up. It's interesting that the writer addresses this question in the scene between Billy and Ted. Billy thinks his mommy left because of him. Ted explains that she didn't; Ted is the problem, not Billy.

In Act Two, Joanna returns to the story. We see her watching Billy from a distance, at a restaurant near his school. She still loves him and her pain breaks our hearts. She takes Billy out for a day and can't let him go. She desperately loves him. She's gotten her life together and Billy still misses her. Ted tells her to go to hell. She gains more sympathy. A boy needs his mother as well as his father. It's a complicated issue.

Interestingly enough, the subplot involving Margaret and her divorce from Charlie parallels the issue of the main plot. In the first place, it gives us a nice contrast to Ted bringing up Billy. Margaret is a single woman bringing up her kid and doing a better job than Ted. This puts more weight on the side of motherhood. When Ted asks Margaret about taking Charlie back, she makes an interesting comment: "I guess it's different if you don't have children . . . even if we're sleeping with other people, even if Charlie was to marry again . . . he's still my husband . . . till death do you part" (Benton, 1978, p. 48).

It's Margaret's position that Charlie will always be connected to her because they share children. He will always be her husband. In the subplot, Margaret and Charlie get back together; Ted and Joanna do not. The fact that having children really means a kind of marriage for life makes the issue and conflict between Joanna and Ted even more complicated and poignant.

Other characters take different positions on this issue. O'Connor, Ted's boss, takes the traditional position. He thinks the grandparents should be taking care of Billy. A man's primary commitment is to work, not to childrearing. Mrs. Willewska, Ted's housekeeper, supports Ted's position that he should retain custody. Even though the judge rules in her favor, Joanna realizes in the end that Billy should remain with Ted.

She gives Ted custody and, once again, breaks our hearts because we know how much she loves Billy. But we also know how much Ted loves him. **[Exercises 1–4]**

The Paper Chase

James Bridges's *The Paper Chase* (1972) was adapted from the John Osbourne, Jr., book that dealt with the life of a first-year law student at Harvard University and the conflict between this student and the hard-nosed Professor Kingsfield, a professor of contract law. The issue is not abortion, gun control, or police brutality, but education—specifically, the question of the value of an education in terms of job marketability. Will getting a Harvard law degree—the paper chase—mean a secure future with a top law firm after graduation? In terms of career success, is it better to have a degree from Harvard versus a degree from a junior college or a state university? If a degree from Harvard is an advantage, then how does one get this degree?

This brings up certain secondary issues that relate to the main issue. Will getting top grades guarantee even more success after graduation? If getting high grades is the answer (being in the upper echelon of the class as Hart, the protagonist, describes it), then what is the best strategy to achieve these grades?

The key to a successful social issue lies in showing us the many sides of the issue and making each side believable. The more complex the issue, the more sides you can deal with, the more interest you can generate among your audience. The complexity of the issue holds the audience's attention and moves the story along. It also generates conflict between the characters because each character represents a different, plausible side to your question. An issue also presents the opportunity for character growth, because one character can change his position on the issue and grow and mature. Issues make your story more interesting, more relevant, and add more depth. Let's examine *The Paper Chase* in more detail to see how this process works.

In the fourth scene of the movie, Hart meets Ford, another student at Harvard. Note the use of names to represent personality traits. Hart is a warm, caring guy. Ford is more businesslike; Kingsfield is the ruler of his class. Bell is fat and dull. Anderson is plain and plodding. In Bridges's scene Ford (1972) says,

```
Of course you've got to have the grades . . .
you can't wear a goddamn Harvard sign
around your neck.
```

Hart replies, "You'll get the grades" (p. 13).

Here, we get our initial view of the social issue. Ford tells us that we need good grades to be successful, and Hart agrees with him. Later in the movie, Hart gives us his strategy for getting these grades. On page 26 of Bridges's screenplay, he tells Susan:

```
The class has divided into three sections.
The upper echelon. They throw themselves
into the fray, speak up in class and in the
end the teachers will get to know their
names and they'll receive better grades and
greater recognition. . . . This week I plan
to enter the upper echelon in Kingsfield's
contract law class.
```

Hart is committed to getting high grades. Like countless students, he feels that the way to achieve his goal is by getting the professor to know his name. How many times have you felt that getting to know your professor might give you an edge when it came to doing better in class? This question is a key one for the story. Will making the professor aware of you lead to better grades? We will return to this question later. Incidentally, this question is not even raised in the novel. Here, we have a case of the screenwriter exploring his main issue through a related secondary issue.

In order to get better grades, Hart starts speaking up in class. He becomes Kingsfield's favorite and even volunteers for extra work. One weekend he has to break his date with Susan, Kingsfield's daughter. Susan is another complication added in the screenplay to intensify the protagonist's problem. Can a student balance grades and romance? Susan reacts by breaking off with Hart. In a scene with Hart, she gives her view on the value of a diploma. Picking up a roll of toilet paper in the grocery store, while talking about the Harvard diploma, she says (Bridges, 1972, p. 90):

```
. . . you can take this and stick it in a silver
box with your birth certificate, driver's
license, marriage license, your stock
certificates, and will! . . . I wish you
would flunk! There might be some hope for
you!
```

Susan's at the opposite extreme on the issue of the importance of the Harvard diploma. To her, it has no more value than toilet paper. Getting a degree might make your career a success, but it ruins your personality and character. What good is having a degree from Harvard if it means Hart will lose his warmth and charm?

Dating Kingsfield's daughter threatens the character's goal of remaining in the upper echelon and getting high grades. Kingsfield almost catches Hart sleeping with Susan when he returns home early from a trip to New York. Hart and Susan are sleeping in his bed. Hart sneaks out in the cold without his clothes. Kingsfield remarks to Susan that he hopes it wasn't a law student. The next day in class, Kingsfield, glaring at Hart, calls on the student sitting next to him, behind him, and in front of him. Hart is "freaked out." In the next scene, Hart meets with Susan. Hart, who thinks Kingsfield was concerned about his progress, is now worried because he thinks Kingsfield knows who was sleeping with his daughter. Susan says,

```
Believe me you're just a name on a piece of
paper. A picture on a seating chart, that's
all. Only one of thousands over the past forty
years. (Bridges, 1972, p. 56)
```

Toward the end of the movie, Hart is alone with Kingsfield and he tells Kingsfield how much he liked his class, how much the class has meant to him. Kingsfield thanks him and asks him his name. Susan was right. Kingsfield doesn't remember his students' names. Hart gets an "A" in the class, not because Kingsfield knows Hart; he gets it on his own merit. The screenwriter answers the question about whether knowing your professors will lead to better grades. The answer for this film is NO.

Interestingly, in the subplot involving Susan and Hart, the question of her marriage comes up in the play. In the last scene, she receives her divorce papers and says, "a piece of paper and I'm free" (Bridges, 1972, p. 134). This is an ironic comment that also relates to the social issue. She was free before the paper came. The divorce decree will not make her free to be with Hart; they already have a relationship. Just like a piece of paper from Harvard does not guarantee success as a lawyer. It's what's inside that counts. Once again, the subplot thematically parallels the main plot line.

In a secondary subplot, Hart and Ford study at a hotel for the final exam. The manager tries to evict them. On page 127 of Bridges's screenplay (1972), Hart says:

```
I know that this piece of paper we signed
out there entitles you to kick us out. But
if you do that I'll call the newspapers and
tell them you're operating a dope ring from
here and you won't get any more business.
So get the hell out!
```

Once again, the value of something in writing comes out—in a contract. Although Hart intellectually agrees with the value of a contract, we see that his emotional response is quite different. He is changing and experiencing character growth. He realizes that grades aren't everything when he throws his unopened letter, containing his grades, into the ocean. He has come to his final realization and moment of growth regarding the value of the Harvard law school experience. **[Exercises 5–7]**

The Bridges of Madison County

In the novel version of *The Bridges of Madison County*, the conflict and issue of the story center on the difference in lifestyle between Robert and Francesca. When Robert asks her to go away with him, Francesca tells us (Waller, 1992, p. 115):

> You're old knapsacks and a truck named Harry and jet airplanes to Asia. . . . Don't you see that I love you so much that I cannot think of restraining you for a moment? To do that would be to kill the wild, magnificent animal that is you, and the power would die with it.

Francesca's life, on the other hand, is just the opposite. She says:

> Yes it's boring in its way. My life, that is. It lacks romance, eroticism, dancing in the kitchen to candlelight, and the wonderful feel of a man who knows how to love a woman. . . . But there's this damn sense of responsibility I have. To Richard, to the children. Just my leaving . . . might destroy him. (p. 115)

The issue in this story is the question of responsibility. Robert has no responsibilities. He is totally free and this freedom makes his love for Francesca even more ideal. Francesca is bound by the conventions of society. The question of her commitment to Robert, versus the delusion she lives with Richard, is answered by her speech on responsibility.

In the movie version, the issue of responsibility is maintained but the screenwriter, in his attempts to intensify the conflict between Robert and Francesca, adds the wrinkle of Robert's past into the equation. Francesca says (Bass, 1993, p. 81),

```
These women friends of yours all over the
world. Do you write to them? . . . See them
again? I just need to know your routine.
```

Robert replies that there is no routine. Francesca continues,

```
You have this habit of not needing and it's
very hard to break. I'll have to be wondering
if you'll be talking to some housewife in Romania.
```

In the movie, Francesca still feels a sense of responsibility to her family when she tells us later about how children take one's life with them. But the movie adds this dimension to Francesca. Her fear that her affair with Robert is just another in a series of relationships he has had along the way with other lonely women helps to dimensionalize her character. We see and above all feel her vulnerability. In the novel,

Francesca appears weak for giving up love for responsibility. But in the movie we see that this issue is more complicating than the choice of abandoning one's family for a perfect love. No love is perfect. No character is without her or his own little fears and doubts. **[Exercises 8–9]**

Harold and Maude

Harold and Maude is one of the classic movies from the 1970s. It's an offbeat romance from a newspaper article about a 21-year-old man who marries an 80-year-old woman. Harold is the protagonist, oppressed by his rich mother who ignores him. For revenge, Harold stages fake suicides and goes to funerals. He drives an old hearse and seems fascinated by death.

Maude, Harold's ally, is into life. She likes to steal cars, hassle the police, and visit funerals. The basic issue of this story is life versus death. Harold is into death as symbolized by his fake suicides. Maude is into life. In Higgins's movie (1971), she visits funerals because it's exciting to her: "The end and the beginning. . . . The great circle of life" (p. 21). Harold wants to escape life. Maude steals cars because she likes to have new experiences. When Harold tells her this is upsetting people, she says, "It's just a gentle reminder not to hold on to things. Here today . . . gone tomorrow" (p. 23). She's into living life to its fullest. "Live," she tells Harold, "otherwise you don't have anything to talk about in the locker room" (pp. 63–64).

The other characters in the story have their views on how one should live life. Harold's mother thinks he is living a frivolous life, the life of a child. Her answer is to settle down and get married. When that doesn't work, she instructs Harold's Uncle Victor, a one-armed soldier, to induct him into the Army. Victor's view on life is an interesting one. He thinks the Army is the answer. Sure there are disadvantages, he tells Harold, while ironically touching his missing arm, but it's a great life . . . travel, adventure, killing.

In one respect, the message of this movie is similar to the one in *The Graduate*. What should a young man do with his life? The answer in *Harold and Maude* is best expressed through the Cat Stevens' song that plays in the background. As the song says, "You can be anything you really want to be." In an interesting cemetery scene, Maude asks Harold what kind of flower he would like to be. He picks a sunflower because they're all alike. He wants to live his life in anonymity, be like everyone else and escape from life. But Maude points out that the flowers are really different from one another. As she says, ". . . part of the world's problem comes from people who want to be like this, but are really like that" (Higgins, 1971, p. 40). This dialogue is superimposed over a visual shot of thousands of graves in a cemetery. People get their chance to be like everyone else after they die, so why not take advantage of life now and "be what you want to be"?

Harold learns this lesson at the end of the movie and he experiences character growth. After Maude dies, it looks like he's so upset that he's going to kill himself. In the climax scene, his car speeds toward a cliff and goes over the edge, but Harold's not in the car. Instead, he's on the top of the cliff playing music with his banjo, "The Cosmic Dance." Harold has finally chosen life over death. By the resolution of the story, we see how the writer feels about its social issue. As long as you're alive, you can always start over, find a new friend, make music. **[Exercises 10–12]**

Roe versus Wade

Nowhere can the concept of social issue be more readily demonstrated than in a typical TV movie. It almost seems that every social issue of the week becomes a TV movie. One of the better TV movies is Alison Cross's (1989) *Roe versus Wade*.

Basically, the story deals with a poor Texas woman who becomes pregnant and cannot afford to raise her child. It's interesting because all aspects of the abortion issue are raised. The shame a woman goes through when she is unmarried and becomes pregnant is best expressed through the mother when she tells Ellen, the protagonist, "you've got the morals of a cat" (p. 20).

On the other side of the issue is Ellen's father. His first reaction is typical, "Does he know?" (p. 24). It's the man's responsibility; let him take care of the problem—the traditional answer. When Ellen replies that his knowing won't change the fact that she's pregnant, the father comes back with one alternative that is pushed throughout the movie, "Are you going to give it up?" Ellen's position is clear:

```
I've got no money. I've got no place to
go. . . . I can't have another baby.
```

But she has a conflict. She can't have another baby and she can't give it up. Why?

```
What can it possibly be like to have a kid
of yours out there getting his butt kicked
and you don't even know about it? (Cross,
1989, p. 26)
```

This is her dilemma. Her problem is that abortions in Texas are illegal. Who's right or wrong? Is she a tramp? Should the father of the child take responsibility? Should she just give the child up for adoption? These are all legitimate questions within the framework of this complicated issue.

In the next scene, Ellen visits a doctor. She tries to make her case for an abortion more demanding by claiming to have been raped. Now we get the view of the medical community on the issue of abortion. She tells the doctor she doesn't have any money and she's sleeping in the back of a truck. His position—that the only abortion allowed in Texas is to save the life the mother—doesn't help her. He refuses to break the law and suggests that she give up the child. But as Ellen points out, "It's the hardest thing in the world to give up a child" (Cross, 1989, p. 27).

What about an illegal abortion? In the next scene, the writer answers this question. Ellen visits an illegal abortion clinic. The audience, as well as Ellen, is appalled by the condition of the clinic. The primitive equipment and unsanitary conditions represent a real threat to the mother's life. The answer is specific. Illegal abortions are dangerous. Once again Ellen is in a dilemma:

```
I either get caught on some table and get
cut up or I have to carry the baby for nine
months and then I have to give it up. (p. 29)
```

In desperation, Ellen consults two female lawyers. The scene takes place in a local restaurant. Most of the dialogue supports the right of a woman to have an abortion (Cross, 1989, p. 33):

```
                 LAWYER 1
Like whether or not a woman's got the right
to make up her own mind and not just in
case of rape or if your life is in danger.

                 LAWYER 2
. . . contraceptives fail . . . your luck runs
out!

                 LAWYER 1
If you can't take care of a baby it shouldn't
matter what the reason is.

                 LAWYER 2
. . . you shouldn't have to have a child and give
it up to strangers just because the state
of Texas says so.
```

Here's one side of the legal position—the right of privacy, the right of a woman to control her own body. But what's the other side? We quickly find out in the next scene (Cross, 1989, p. 36) between Lawyer 1 and the law professor:

```
                    PROFESSOR
Call it protoplasm if you like . . . it's
the end of a life.

                    LAWYER 1
Women are dying. That's the end of life.
You know as well as I do that the law
doesn't stop abortions. It just makes them
dangerous.
```

Note the comeback line by Lawyer 1. She has the tag line in the scene, the final line that is hard to top. Of course, this seems to be the writer's position, so the cards are really stacked against the professor's having a good comeback line.

Finally, the case goes to court. In the lower courts, the characters' arguments further develop the different sides of the issue. The Assistant District Attorney feels that the embryo is a human at the moment of conception and, thus, abortion is murder. The other side argues that a fetus isn't even a person under Texas law; it has no rights. Meanwhile, the rights of the mother are being violated. The screenwriter cleverly gives us every side to this issue without really stating his own views. In a way, each side's argument has a certain validity.

Since the production of this movie, abortion laws in Texas and throughout the country have changed. A writer using this issue today would have to take these changes into account and develop new arguments that reflect the liberalization of abortion laws. The question of keeping or giving up the baby remains the same but the question of the health of the mother might be changed, for example, by making the protagonist further along in her pregnancy. Second- and third-trimester abortions are still very controversial. Congress has repeatedly tried to outlaw late-term abortions. **[Exercises 13–15]**

SUMMARY

Social issues make a good foundation to any dramatic story. They give your story added depth and seriousness, and set up a natural series of conflicts. The key to a successful issue lies, however, in subtlety. The audience doesn't want to hear characters expounding on their philosophy of life. They want an interesting story that deals with something of substance. The story can be serious or it can be humorous, but above all it should make the audience think, without being obvious. Whether one is adapting or modifying an existing social issue from a novel or short story, or creating an original idea with an interesting issue, the social issue is an important dramatic device for enhancing the quality of any screenplay.

EXERCISES

1. In the courtroom scene, Ted gives a long, emotional speech supporting his position that he should retain custody of Billy. Write his speech, as well as speeches from Joanna and Billy, giving their positions. Then, check the appendix for a summary of these speeches.
2. Imagine that you and your boyfriend or girlfriend are planning to get married but have different religious backgrounds. Using conflict, discuss how the child should be raised. Give reasons to support your position. What would your priest, rabbi, or minister have to say? Write his or her speech. Have your instructor check the quality of these speeches.
3. Now that you've been married for a few years, you realize that the problems of your marriage cannot be overcome. Yet, you have a small six-year-old child at home. What do you do? Do you stay in the marriage, separate, or divorce? What about custody of the child? Write a scene in which you discuss these issues. Now change the sex of the child. Rewrite the scene. Does this change make a difference?
4. Consider an interracial marriage between Sarah Schwartz and Leroy Jones, two fictitious characters. How would they deal with a custody case involving little Loretta and Brad? Do a brief character sketch and write the scene.
5. For Hart to experience character growth, he must gradually come to the realization that grades are unimportant. One night, Hart visits his friend from school, Brooks, who is having trouble in Kingsfield's class. He brings him some notes. Hart finds out that Brooks's wife is pregnant. Brooks is drinking heavily, and says, "with my grades it's going to be hard keeping us in pabulum" (p. 104). Hart replies, "They're just grades, Kevin." Here is our first indication that Hart is changing his position. Kevin responds vehemently on the other side of the issue telling us just why grades are important. Write this speech and keep it short and emotional.
6. Write Hart's rejoinder to Brooks in keeping with his new view on grades. Keep your dialogue connected. In other words, respond to some key phrase in Brooks's speech in your comeback line. See Chapter 10 for more information on dialogue.
7. Write your own views on grades in a short scene between yourself and your parents, a friend, or a teacher. Set the scene and be sure to generate conflict by taking opposite points on the issue. For example,

you might have a scene between yourself and a teacher in which you're debating on the grade you received in her course last semester. Bring up your issue on the importance of grades, but be subtle.

8. Write an emotional breakup speech from a man's point of view and then from a woman's point of view.

9. Write this speech from the position of a friend of either the man or woman.

10. Use the fictitious characters of Susan Alcott and Henry Chapman II to discuss abortion, wife beating, and drug abuse. First, describe these characters and then write a scene with them. Make sure that both sides of the issue are represented.

11. In a *Harold and Maude* scene set in Maude's apartment, Harold tells Maude he's never lived, but he's died a few times. They both go on to tell backstories about themselves that illustrate their views on life and death, the issue of the movie. How do you think this scene should be written? Compare your results with the real version in the appendix.

12. Watch several old movies on TV. See if you can pick out the social issue of each. Listen for different sides of the issue. Which characters represent which side of the issue? What is your view on the issue? What is the writer's real viewpoint and how is this determined? How does the subplot illustrate the theme of the main plot?

13. The issue finally reaches the Supreme Court. Lawyers for and against the issue argue their case in a brief soliloquy before the court. Write their speeches, developing both sides of the argument. Make sure the jargon sounds legal, but keep the tone emotional.

14. Abortion is a volatile issue. Devise a scene between a pro-lifer and a pro-choicer. Develop conflict and find some new aspect of the issue to argue that has not been covered in the previous information.

15. Pick another issue, such as drugs or alcoholism. Outline the different points of view. If you were dramatizing a story, what characters would you create to represent the different views on your issue?

7

Action Adventure and Science Fiction

Jurassic Park and *Apollo 13*

Action adventure and science fiction have always been popular sources for movies since the days of *Ben Hur* in the 1920s. In recent years, we have seen many popular adaptations along these lines—*The Silence of the Lambs, The Firm,* and *Hunt for Red October.*

JURASSIC PARK

One of the most popular and successful action adventure movies in recent years is Michael Crichton's *Jurassic Park* (1990). Like other adaptations, Crichton's adaptation of his own work follows many of the same patterns we have seen for other novels.

Starting Point

The novel begins on the island of Costa Rica (Crichton, 1990, p. 1):

> The tropical rain fell in drenching sheets, hammering the corrugated roof of the clinic building, roaring down the metal gutters,

> splashing on the ground in a torrent. Roberta Carter sighed and stared out the window. From the clinic, she could hardly see the beach or the ocean beyond cloaked in fog. This wasn't what she had expected when she had come to the fishing village of Bahia Anasco, on the west coast of Costa Rica, to spend two months as a visiting physician.

The novel's opening scene introduces Roberta Carter, a character totally omitted from the movie. Then, the arrival of a worker mauled in a construction accident and in need of immediate treatment suggests some future problem on the island: "Bobbie was thinking about the boy's hands. They had been covered with cuts and bruises, in the characteristic pattern of defense wounds" (p. 6).

The movie, on the other hand, opens with the workers on the island watching the arrival of a huge crate which we find out later contains the deadly pack-hunting raptors capable of slashing their victims with deadly razor sharp claws. Visually, the scene works nicely. The bushes are swaying and we hear horrible animal sounds. Suddenly, one of the workmen is seized from inside the cage by the raptor. Another workman, along with the chief of animal control, tries shocking the animal with electric stun guns. The raptor slowly pulls its victim inside the cage as we see the words "Jurassic Park" written on the hard hat of one of the shocked men.

Although the opening incident is the same in both the screenplay and the novel, the approach to presenting the material is different. With their introspective approach to storytelling, novels take a more detailed look at a particular incident. The characters explain their feelings as they go along. Bobbie talks of her surprise at the weather conditions in Costa Rica. The focus on the incident with the worker is approached more from Bobbie's reaction to the strange wounds than from the incident itself . . . slowly building up suspense for the rest of the novel.

In the movie, the material is presented from a visual point of view in the most economic and quickest fashion. We establish where we are, for example, by the use of the name on the hard hat rather than through dialogue. Presenting the raptors in the opening scene is also effective because they are the central problem in the movie and the opening scene is therefore used to dramatize the intensity of the opposition the protagonist is to face in the story. A strong visual opening, showing the accident rather than picking up the story after the accident as in the novel, is a better hook for the audience and starts the story off with a bang. Now that the audience is hooked, the more expository details, such as the introduction of the protagonist and other characters in the story, can proceed. **[Exercises 1–2]**

Structure

Chapter 2 of the novel is called "Almost Paradise" and begins with a couple and their young daughter driving in an isolated section of Costa Rica looking for a deserted but beautiful section of the beach. When they arrive at the beach, their young daughter wanders off. "I feel so isolated here," Ellen said. "I thought that's what you wanted," Mike Bowman said. "I did." "Well then. What's the problem?" "I just wish I could see her" (Crichton, 1990, p. 15). If she could have, what she would have seen was her daughter being attacked by several small lizard creatures. This scene becomes the opening scene in *The Lost World*, the sequel to *Jurassic Park*. Because the screenwriter is forced to condense 399 pages of novel into 120 pages of screenplay, incidents, characters, scenes, and even entire chapters of the novel often must be cut.

Even though the screenwriter attempts to remain faithful to the intent of the novel, duplication of the structure of the novel is often impossible or likely to prove unsuccessful. This means making cuts. Fortunately in the case of *Jurassic Park,* some of this lost material has found its way into the sequel. Students often ask me if the writer plans this upfront. My feeling is no. There's no way to anticipate the success of a movie. However, the original screenplay for *Back to the Future* was written with the idea of successive movies and even filmed with this in mind. But, for *Jurassic Park,* the screenwriter went back to his original source material to pick the material that formed its sequel. **[Exercises 3–4]**

Introduction of the Protagonist

Chapters 3 and 4 of *Jurassic Park* follow up on the incident on the beach. Marty Gutierezi, another character omitted from the movie, searches the area of the attack for clues to what happened to the little girl. A strange lizard is discovered, which leads Marty to think he has discovered a new species. In Chapter 5, the remains of Marty's discovery are shipped to Columbia University Medical Center to the lab of Dr. Richard Stone and his assistant Alice Levin, two additional characters omitted from the movie. Alice suspects the creature is a dinosaur, not a lizard, but the general consensus is that such a probability is impossible. Finally, on page 31 of Chapter 6, we meet the protagonist Alan Grant at his archeology dig sight in the desert.

Although the introduction of Grant remains the same in both the novel and the screenplay, as one would expect, major information is omitted for the screenplay adaptation. On page 37 (Crichton, 1990), for example, we learn:

> Alan Grant had found the first clutch of dinosaur eggs in Montana in 1979 and many more in the next two years but hadn't gotten around to publishing findings until 1983. His paper with his report of a herd of ten thousand duck-billed dinosaurs living along the shore of a vast inland sea building communal nests of eggs in the mud, raising their infant dinosaurs in the herd made Grant a celebrity overnight.

Novels contain vast amounts of backstory not only about the protagonist but on most of the major characters in the movie. In the novel, for example, we learn that "Hammond was flamboyant . . . a born showman and back in 1983 had an elephant that he carried around with him in a little cage. The elephant was nine inches high" (p. 59). Again the screenwriter must omit and condense the vast amounts of backstory in the novel as he writes the screenplay. The best approach is to focus on the main characters and decide which backstory information is most relevant to the story.

Often the character has a traumatic incident which haunts him in the story. We call this backstory the character's ghost: something that happened before the movie begins but continues to influence the character's actions during the story.

In *The Silence of the Lambs*, for example, Clarise is haunted by the death of her father and the incident with the slaughtered lambs when she lived for a brief time on a ranch in Montana. Bits and pieces of this backstory should be given to the audience throughout the movie with the actual incident revealed somewhere in the third act. Not only does this hold the audience's attention, but it serves to dimensionalize your characters.

In Chapter 7, we meet Ellie Grant, the other protagonist of the story. Again an entire chapter of the novel is devoted to the introduction of this one character. We see Ellie at work on bones in an acid bath. In the movie, both Ellie and Alan are introduced in one scene. This picks up the pace and helps move the story along. In the novel, Hammond, the character responsible for Jurassic Park, contacts Ellie and Alan by phone to invite them to Jurassic Park. A few days later, plans of the park are sent to them, and Alan and Ellie discuss the elaborate fortifications surrounding the park. In the movie, this sequence is changed. In an effort to speed the story along and also to add more visual effect, Hammond arrives in the scene following the audience's introduction to Ellie and Alan.

At this time we are also introduced to Alan's knowledge about dinosaurs and his dislike of children, which sets up the character-growth segment of the story since Alan will be interacting with Hammond's grandchildren during the course of the movie. Hammond's arrival by helicopter in the next scene is a much more effective introduction than the telephone conversation in the novel. Telephone conversations in movies can be really boring. Whenever possible put your characters together to confront each other; it makes for more interesting drama. Hammond's arrival is also much more visual because the dust blows over the remains of the raptor skeleton that has been discovered. It makes a symbolic statement for the movie as the dead past is wiped away by the technology of the present. Ironically, technology cannot control this past as the dinosaurs escape on the island. **[Exercises 5–6]**

Condensing Scenes

The lengthy nature of the novel compared to the relatively short time span of the screenplay necessitates the difficult decision of what material to eliminate for the screenplay. This choice should be made primarily on the basis of keeping the story moving in a straight line while holding the audience's attention. Character development in terms of backstory and scenes that are used primarily to explain character motivation must be eliminated. Use of flashbacks should be kept to a minimum. In *The Silence of the Lambs,* flashbacks are used effectively to reveal the protagonist's ghost because they are given to us in brief scenes that do not impede the forward movement of the story.

One possible solution is to combine scenes. The novel *Jurassic Park* devotes separate scenes to the development of the Donald Genero character and the meetings with the board of directors that funds Hammond's island. When we first meet Genero in the novel, for example, he is in his law office in San Francisco discussing Hammond: "We can't trust Hammond anymore. He's under too much pressure. The EPA's investigating him" (Crichton, 1990, p. 49). Instead, in the movie, this character is introduced for the first time when he arrives on the island, and we realize he's there to check up on Hammond. Thus, the board of directors meeting with Genero, which forms all of Chapter 11, is omitted. The story moves along quickly.

This same economy of movement can also be seen in the introduction of the mathematician, Malcolm. In the novel, we meet him on a plane from Dallas on his way to the helicopter ride to the island. In the movie, this meeting is speeded up and we see him for the first time in the helicopter. Even the introduction of the first sightings of dinosaurs on the island has been changed from the book. In the novel, Grant sees a small movement, which he mistakes for a tree. It is, of course, a dinosaur. In the movie, Grant sees an entire herd moving across an opening plain. It is much more visual.

Malcolm, as a mathematician, espouses an entire theory of life, which he calls chaos theory (Crichton, 1990, p. 41):

> What we call nature is in fact a complex system far greater than we are willing to accept. We make a simplified image of nature and then we botch it up. . . . We build the Aswan Dam and claim it is going to revitalize the country. Instead it destroys the fertile Nile delta, produces parasitic infection and wrecks the Egyptian economy.

Later, on page 171, he tells us:

> And that's how things are. You start out doing one thing but end up doing something else, plan to run an errand but never get there. And at the end of your life, your whole existence has that same haphazard quality to you too. Your whole life has the same shape as a single day.

In the movie, Malcolm retains his philosophy on the chaos of life but it is reduced to two or three key scenes, with his speeches greatly reduced. **[Exercises 7 and 8]**

In addition, information on Malcolm's previous marriage, Hammond's recruitment of the geneticists responsible for life on the island, and other key scenes in the novel are cut. For example, in the novel, there are two Rexs, a large one and a baby one. The baby and his battle with the big Rex are cut. In the novel, the big Rex pursues Alan and the children across the island. Alan and the children spend some time in a boat on a river trying to get back to the main building—the inspiration for the Jurassic Park ride at Universal Studios. Along the way they are stalked by the Rex and also by flying reptiles who attack them in an aviary along the way. The scene of the Rex attacking them in the waterfall makes its way into the movie's sequel, *The Lost World*. The attack of the dilophosaurs from the novel is also omitted.

Change in Characters

Hammond's grandchildren have been changed in the movie. The girl is made more capable and computer literate than she is in the novel, in which she is basically helpless. This change ties in with the change in Ellie's character. Women characters are made stronger and less of a cliché, which is a change that appeals more to current audiences. Ellie's statements in the movie about the equality of women have also been added.

Grant is also different in the movie. Because he is the protagonist, scenes are changed to involve him in the story more. Remember, in a movie the protagonist must be seen in action and should be involved in almost all major scenes. Another change involves the incident of Timmy being trapped in the tree. In the novel, Timmy, who is the more competent of Hammond's two grandchildren, rescues himself from his dilemma. In the movie, Grant climbs the tree to bring him down. We see the character (protagonist) in action.

Change in Ending

Picking the right ending moment for a movie is as crucial as finding the proper starting point. In the screenplay, the action must build to a climax followed by a short resolution scene as the movie ends. In the novel, Hammond and Malcolm are both killed. Grant kills the raptors and the Costa Rican military arrives to wipe out the island and the animals on it. Grant also tracks down the breeding place of all the animals. In the movie, the raptors are killed, one by Grant but the others by the Rex. This change supports the theme embodied in Malcolm's theory of chaos—the unexpected occurs to save the day. Normally the protagonist solves the problem of story, but this ending is effective because it ties in with the theme of the movie.

The deaths of Hammond and Malcolm have been eliminated from the movie, thus affording them an opportunity to return for the sequel. In the movie, the ending occurs with the death of the raptors and the departure of the characters from the island. The search for the nesting place and the arrival and burning of the island by the soldiers are really anticlimatic. Remember, the main problem throughout the movie has been the raptors and solving this problem really marks the ending of the story from a dramatic point of view.

Again time consideration is another factor that must be taken into account. Not everything of interest from the novel can make its way into the movie: "Somewhere behind them they heard explosions and then ahead they saw another helicopter wheeling thru the mist over the visitor center and a moment later the building burst in a bright orange fireball" (Crichton, 1990, p. 396). In the sequel, the characters visit the still-standing visitor's center. **[Exercises 9–10]**

APOLLO 13

"One small step for man; one giant step for mankind."

Starting Point

Like other action-oriented novels, *Apollo 13* makes the transition from 398 pages of narration to 120 pages of screenplay through both condensation and omission of key scenes, characters, and events. The novel begins with a prologue about the use of poison pills in an emergency (Lovell and Kluger, 1995, p. 1):

> Pills got started. Most people had heard them. Most people even believed them. The press and the public certainly did . . . stories about the poison pills always made Jim Lovell laugh. . . . There just weren't any situations in which you'd ever really consider making, well, an early exit.

Chapter 1 begins with Lovell at the White House having dinner and doing what was called his time in the barrel: "Lovell was not unaccustomed to this kind of evening. . . . This was just more of what he and the other members of the astronaut corps called their time in the barrel" (p. 8). This chapter also talks about President Johnson, the treaty about returning lost astronauts to their home country, the picking of the astronaut delegation for the night's dinner, Lovell's talk with the Soviet ambassador, the differences between the various spacecrafts—the Mercury and Gemini versus the Apollo—and then the fire aboard the Apollo training mission. The chapter concludes with how each of the astronauts received news about the accident, the astronauts getting together and drinking, and the funeral at Arlington.

In the movie, the initial focus is on the tragedy of the fire. All the other material in the prologue and Chapter 1 has been eliminated. Instead, the screenwriter uses the fire as a starting point for a brief narrated history over the opening credits of the Apollo program leading to the introduction of the protagonist, Lovell, driving home in his red Corvette (not in the book) as we listen to the first landing on the moon. Lovell arrives home while the first moon walk is taking place on TV. At this time, we meet three of the other key characters in the movie—Marilyn, Lovell's wife; Swigert, one of the pilots on the Apollo 13 flight; and Ken Mattingly, who loses his chance to make the flight because he was exposed to measles.

In the novel, the introduction of Swigert, played by Kevin Bacon in the movie, includes this reference to his single status (Lovell & Kluger, 1995, p. 88):

> The tall, crew cut Swigert had the reputation of a rambunctious bachelor with an active social life. Whether this was true or not was unknown. . . . His Houston apartment included a fur-covered recliner, a beer spigot in the kitchen, winemaking equipment and a state-of-the-art stereo system.

In the movie, in which the writer stresses the visual over the narrative, we see Swigert in Lovell's house flirting with a woman as they discuss the docking of two spacecrafts, with its obvious sexual undertones.

The third scene of the movie puts Lovell outside with his wife, Marilyn, discussing the moon and Armstrong's new place in history. Marilyn asks Lovell to show her the location of the mountain named after her on the moon. This scene does not exist in the novel. It is added to show the relationship between Lovell and Marilyn as they play on the chairs in the backyard. In the novel, the writer describes Lovell's naming of Mt. Marilyn on his first flight to the moon. With the focus on the upcoming flight of Apollo 13, this backstory is omitted from the movie. **[Exercises 11–12]**

Elimination of Scenes and Backstory

Chapter 2 of the novel details Lovell's previous flight on Apollo 8. We get backstory on the members of that crew such as Borman. One of the Apollo astronauts is lost in a plane crash and Lovell is assigned to the investigation team. Borman spends time lobbying for changes in the Apollo craft. Problems with the LEM are discussed with the bulk of the chapter devoted to the launching and mission of Apollo 8.

Chapter 3 takes us back in time. We learn of Lovell's interest in space which began in his teen years, his decision to join the Navy, and his years in college and at the academy. We are given details of his first meeting with Marilyn and their subsequent courtship. Lovell's flight training after graduation is followed by an incident on an aircraft carrier in which Lovell is almost killed while trying to land his plane at night—he saves himself by cleverly following the ship's wake in the dark.

All the information in these two chapters is omitted from the movie. Interestingly enough, the backstory about Lovell's aircraft carrier adventure does survive into the movie but not until the third act where the screenwriter uses it to show us the "stuff" Lovell is made of in emergency situations. This revelation adds to the tension as Lovell faces the greatest emergency of his life, the return of his crippled spacecraft to earth.

Chapter 4 deals with the character of Sy Liebergot, one of the controllers in Houston. He is given a test situation which he fails during a routine training exercise for the coming Apollo 13 flight. Chapter 5 deals with the character of Wally Schirra who makes a brief appearance early in the movie. We also meet the character of Chris Craft who is omitted from the movie. Backstory about Aaron, another character in this chapter, is also omitted.

Additional backstory about Lovell, which appears throughout the novel, is also omitted from the screenplay; the writer judiciously decided which backstory is relevant enough to find its way into the movie. We are told of Lovell's experiences at the Navy Test Center, his graduation at the top of his class, and his disappointment at not being assigned as a test pilot. A lot of information is also given to us regarding Lovell's attempts at becoming an astronaut, his failure during the first round of admissions, and his medical problems. Lovell's interview about the risks of space travel compared to the war in Vietnam is also discussed.

Scenes in the novel involving detailed descriptions of the LEM, including its history, have also been omitted. Minor characters have also been cut. We lose, for example, Father Raish, who comes to the Lovell's house in the novel to comfort Marilyn. We also lose Mel Richmond who is in charge of the recovery team and Don Arabian, known as Mad Don in the novel. Interestingly enough, President Nixon, who is a figure in the novel, is eliminated from the movie. During the crisis, he calls Marilyn on the phone and is given constant updates on the progress of the crew. During the recovery phase, Nixon is present on the recovery vessel but steps aside so Marilyn can be reunited with her husband. Nixon never appears in the movie. **[Exercises 13 and 14]**

Adding Scenes

For dramatic purposes, a writer often embellishes and even adds material that is not included in the novel to dramatize a particular moment, create tension, or add additional conflict. In *Apollo 13,* we have several of these moments. Prior to take-off on Apollo 13, Lovell has a dream of a crash that is added in the movie. The scene with him talking to his son prior to take-off, which rounds him out as a character, is also added. Marilyn, who is growing increasingly apprehensive about the flight, does not go to the launch in the movie. This adds conflict and tension to the scene. Marilyn's loss of her wedding ring before the flight is also added and symbolizes some possible future problem. During the launch itself, we see Ken Mattingly watching miserably from a distance as the ship takes off. As the pilot bumped from the flight, this little bit of drama is added to the movie to reveal Mattingly's disappointment.

On the flight itself, several key elements have been added to the movie. Haise throwing up and the initial glitch with the lift-off rocket foreshadow future problems with the men and equipment on the flight. The problem of Swigert docking the ship is also added for tension. "If he doesn't do it we don't have a mission," helps to add suspense as well as increase the audience's doubt about his abilities as a pilot. Later conflict is added in the scene when Haise accuses Swigert of doing something wrong during the cryo mix: "All I did was stir these tanks. This is not my fault."

The carbon dioxide problem involving the difficulty of getting rid of the recycled gas is in the novel but it is "juiced up" for dramatic purposes in the movie. The problem is solved in both novel and screenplay but in the movie there is more urgency . . . more of a time problem and even the moment of trying out the new apparatus has been dramatized to add tension. Haise rips the first plastic bag and, with time running out, a new bag has to be constructed. The problem with the manual burns is another example of a scene from the novel that has been altered to add tension. In the novel, the characters have to execute a difficult manual burn in order to put their spacecraft back on the proper reentry track. In the movie, this moment has been intensified with the men having a difficult time executing the maneuver in the allotted time.

Additional conflict is added through the character of Gene, the mission control director. With the elimination of the Chris Craft character, Gene drives the men to come up with solutions in a much more dramatic fashion than we find in the novel: "No estimate goddamn it . . . I want the procedure now." Later he even argues over the importance of the mission. "Worst disaster NASA ever experienced . . . with all due respect, sir, NASA's finest hour."

Climax and Resolution

Several elements are added in the climax scene to create more tension—the scenes with Mattingly trying to figure out the exact amount of electrical requirements for a successful return. The introduction of a possible typhoon in the landing area is another complication that wasn't in the novel, plus the danger that there may be some problem with the parachute deploying properly is added.

In the movie, the ending and resolution occur following the men's safe return to earth. Like *Jurassic Park*, the *Apollo 13* novel concludes with a chapter that details the investigation of the accident on the spacecraft. We learn the results of that investigation and what happened to the men on the spacecraft in the years following their return to earth. In *Jurassic Park*, the final chapter was given over to the clean-up of the island. For the screenwriter, finding the proper ending point is just as important as finding the proper starting point. Stories that drag on past the proper ending point can ruin the writer's attempt at building and resolving the tension that's been created in Act Three. **[Exercises 15 through 20]**

SUMMARY

Action adventure and science fiction movies are growing increasingly popular as strides in technology are made. But these types of movies often sacrifice character development and plot for action. Above all, movies, such as *Jurassic Park* and *Apollo 13*, must hold their audiences' attention. Starting and ending points are critical. The story needs a hook to grab us and must end at the proper point to avoid the pitfalls of lengthy and boring exposition.

Characters must be eliminated, backstory cut down, and the story moved forward. For a story like *Apollo 13*, these requirements can be complicated by the truth of history itself. When the writer cannot make up events, as for *Jurassic Park*, he or she can at least stretch the truth about certain moments to add drama and tension to an event. And, even in the midst of all the action, the screenwriter can still focus on showing the protagonist's character growth and deal with a relevant issue—the misuse of technology in *Jurassic Park*, or triumph over technology in *Apollo 13*.

EXERCISES

1. Describe two other action-oriented movies that use an action sequence to hook the audience.
2. Pick a monster—a creature from outer space, a science experiment gone wrong, or a serial killer-psychotic type—and create your own action-oriented opening scene.
3. Pick a novel and film combination and list the scenes from the novel that were cut when the screenplay was written.
4. Examine a novel, such as *Jaws*, that ultimately produced a sequel and determine if any of the material in the original novel appears in the sequel.
5. Write a telephone conversation for two characters and then rewrite the scene putting the same characters in the same location. Which one is better and why?
6. Find a telephone conversation in a movie. Does it work? Why?
7. Take a character's philosophy as it is presented in a novel and condense it down to one or two lines of dialogue.
8. In two contemporary movies, find examples of characters who state their philosophy in dialogue.
9. Examine the endings of several movies. Discuss their effectiveness in terms of reaching a climax and a quick resolution of the story.
10. Rewrite the ending of *Jurassic Park* to incorporate more of the material in the ending of the novel.
11. Rewrite the beginning of *Apollo 13* to give us a more dramatic hook opening.
12. Discuss the successes or failures of other science fiction movie openings.

13. Look at key backstory information given in various movies. Discuss how this information is necessary for the understanding of the character.

14. Make up some backstory information on a character of your choice. Write a key scene in which you introduce this information.

15. Look at several movies that have successful endings. Discuss the appropriate ending point of each movie.

16. Look at several movies that have unsuccessful endings. Discuss why these endings don't work.

17. Find the climax scenes of two movies. Compare the climax to the climax in the short story or novel on which they are based. Have any scenes been added or moments "juiced up"?

18. Write an alternative ending, which is negative or ambiguous, for *Apollo 13*.

19. Write a scene for *Apollo 13* in which the protagonist loses "his cool."

20. Discuss other movies by Ron Howard, the director of *Apollo 13*. Are these movies similar? In what ways are they different?

8

The Short Story

There are certain reasons writers prefer adapting from a short story rather than from a novel. The rights to a short story may be easier to obtain than the rights to a full-blown novel. The short story is, by definition, shorter and has less characterization and subplots to deal with, which, for some students, makes it an easier source for adaptation. Those students who do pick the short story invariably run into difficulty because of the length of source material—they often run out of scenes to add into the story or ideas about how to develop the plot. The problem here, I think, is the problem one has with adaptation in general.

In the quest to remain faithful to the original work, which is the number one obsession of potential screenwriters, those who pick a novel find that they have enough material with which to work. This gives them more security. Those who adapt from the short story have to rely more on their imaginations when additional characters and plot details are needed. If students could overcome their fear of using their imaginations to enhance their short story sources, the experience of adapting would be more productive.

Adaptation of the short story into a screenplay presents many of the same problems one encounters when adapting from a novel. Making an adaptation that contains the dramatic components previously discussed in this book should always be the prime consideration when adapting from a short story. Like novel adaptation, short story adaptation needs to consider the taste of the mass audience. The writer needs sympathetic protagonists, well-rounded antagonists, interesting love interests, updated social issues, and happy endings.

ADAPTING FROM THE CLASSICS

One source for short story adaptations has always been classic writers, such as Henry James, Stephen Crane, Edgar Allan Poe, and D. H. Lawrence. In Truffaut's film *The Green Room*, based on a Henry James short story, "The Altar of the Dead," we have the standard problem of adapting from a classical short story—the need to update the material for a contemporary movie audience. As Neil Sinyard (1986) says in *Filming Literature*: "The spirit of James is elusive, distilled as it is in a sensibility and style essentially attuned to an era before the film age" (p. 43). The recent failure of the movie adaptation of Lawrence's *Portrait of a Lady* attests to this difficulty.

"The Altar of the Dead"

"The Altar of the Dead" is basically the parable of a man obsessed with death and the loss of his dead fiancée. In the adaptation, the writer makes several changes. First, he updates the time period from the late nineteenth century to after World War I. This change makes the story more accessible to modern film audiences and also serves to make the protagonist more sympathetic than the cold and dark figure of the short story. The protagonist has served in the war. His obsession comes from this very negative experience of seeing too much death. We sympathize with this man and understand the pain of the loss of his wife early in the film. Note the writer changes the woman from a fiancée to a wife, again to make this obsession more acceptable to the moral conventions of a modern audience.

The adapter also adds plot details and characters. In the film, the character of the housekeeper's boy is added. Julien, the protagonist, takes a liking to the boy and tries to help him overcome his speech impediment. The addition of this character not only adds a subplot and lengthens the film version, but it also serves to make the protagonist more likable because he has also found a way to embrace life through this new character. This relationship also serves as the beginning of the protagonist's turning away from his obsession with death and returning to an appreciation of life. The character experiences growth.

"The Blue Hotel"

"The Blue Hotel" by Stephen Crane is another example of a classic short story that needs to be updated and changed to work as a film. The adapter adds material to his screenplay to increase its length and make it more dramatic.

The main character of the Swede is given backstory in the film that explains his motivations and at the same time makes him more dimensional than the sketchy character in the short story. This backstory also makes the character more sympathetic. The adapter also makes a change in structure. In the short story, the Swede confronts and fights the cowboy for cheating. After winning, the Swede, feeling very cocky, goes off into town where he is killed by another gambler. In the short story, the Easterner reveals the fact that the cowboy really cheated the Swede after the Swede's death. The implication in the short story is that the Swede was destined to die. In the film, the Easterner reveals this information just after the Swede leaves for town. If he had told the Swede before he'd left, then perhaps the Swede would still be alive. As Sinyard (1986, p. 92) says:

> This shift in position of the main events in the narrative has the effect of changing the perspective and tone. Crane's story is one of ironic fatalism. No one is really responsible. . . . Agee [the adapter], however, turns it into a moral tale of courage, cowardice, conscience and collaboration . . . the insistence on the theme of the cowardice of the man who does not speak up against a wrong.

Sinyard makes an interesting observation about the reason for this change. The time period was that of the McCarthy communist witch hunts. Agee was appalled by the lack of people willing to stand up to McCarthy. Thus, his story becomes an indictment against people who don't speak up against wrongdoings. Again, note how the screen adapter changes incidents in the short story (or novel) to update the story into the time period for which the film is written. The same change was made in the adaptation of *Little Big Man*, when the conflict between the Army and the Indians was changed in the screenplay to become a symbol for the war in Vietnam.

"Rear Window"

Even in more modern short stories in which the adapter doesn't have to update the time period of the story, other changes are necessary to make the story conform to a dramatic structure. In "The Rocking-Horse Winner," the negative ending of the short story is changed to a more positive ending for the movie, in keeping with the audience's need for happy endings. In the adaptation of Cornell Woolrich's short story "Rear Window" into the Alfred Hitchcock thriller, the character of Lisa is added to give the main character a love interest, which is necessary for the taste of movie audiences. This character also provides the protagonist, Jeffries (played by James Stewart) with a chance for character growth when he realizes at the end of the film that he is really in love with Lisa.

A film needs other differences to make the plot more dramatic than the short story. A movie opens with a troublesome situation for the protagonist. He really doesn't want to be stuck in his apartment with a broken leg. He has the opportunity to go on another assignment, but because of his leg he cannot leave. This creates a problem and conflict for the protagonist not found in the short story. The character of the boss is also added into the film to give the protagonist more conflict and add scenes to the film. The film also has a certain unity that is lacking in the short story. In the short story, the protagonist solves his problem and is free to leave when the doctor comes to remove the cast on his leg. In the film, the story begins with the protagonist lying in his apartment with a broken leg. It ends with him in the same position, but now he has two broken legs. Films' endings and beginnings are often similar, as if life follows a circular structure.

The Wild One

In the adaptation of Frank Rooney's short story, "Cylist's Raid" into the cult movie *The Wild One*, we find many of the same adapting techniques. The writer simplifies the story by combining his characters. Johnny, the protagonist (Marlon Brando), is a combination of two gang members. In the short story, one of the townspeople is the protagonist, but the adapter makes Johnny his protagonist. The fact that Johnny is really an antihero is in keeping with films of this era. The adapter updates his material. In any case, the adapter still attempts to build sympathy for his protagonist by making him more intelligent than other members of the gang. The love interest with the girl from the town is added to the film version to further add sympathy to Johnny's character, because he is the only member of the gang that really cares about her getting hurt. In the end, the protagonist overcomes his problem with the townspeople and leaves the town with his gang.

There are other differences between the adapting process of the short story and novel. The short story, unlike the novel, is often condensed, with a sketchy plot and few characters. There may be only a few key scenes in the story and little rising action and issue. Therefore, much like the writer of an original piece, the short story adapter is often faced with adding story material such as plot lines, conflict, characters, and subplots, to an adaptation to make the story less narrative and more visual and dramatic.

Later in this chapter, we examine the process of adapting a short story into a screenplay by focusing on three key examples:

1. The adaptation of the short story "It Happened on a Brooklyn Subway" into a 30-minute teleplay.
2. The adaptation of the short story "The Sentinel" into the movie *2001*.

3. The adaptation of the short story "Rita Hayworth and the Shawshank Redemption" into *The Shawshank Redemption*.

First, however, we'll review the problems for students adapting novels and short stories into screenplays.

ADAPTING FROM THE NOVEL

The central problem of adapting from a novel is condensation. The writer needs to decide what plot lines and characters to eliminate from the story to make it work within 120 pages. Making choices is difficult for many beginning adapters because of their desire to remain faithful to the original work.

Recently, a student of mine came to me unhappy about the fact that her adaptation wasn't working. She was adapting a story from a large novel with many subplots and many pages devoted to detailed characterizations. She wanted the characterizations to be part of her screenplay. When I ask her about the purpose of one of her scenes, which seemed wordy and lacked action, she explained the scene was important for character development of the minor characters in the story. This was a scene literally taken right out of the book. Further questioning revealed that her protagonist was only involved in two scenes in her first act with the remaining scenes devoted to her minor characters.

Most students want to include detailed characterization and the wealth of descriptions so predominant in the novel. The screenplay isn't built this way; it needs action that builds during the course of the story to hold the audience's attention and it needs to focus its scenes on the main protagonist and be very selective about what traits it reveals about the main characters. Minor characters who don't push the story along need to be ruthlessly eliminated.

Although it was late in the semester, I suggested that the student literally throw the book away and begin again. She had the basic idea from her source. It was now necessary for her to structure her story as if she were doing an original. I suggested that she create an interesting opening scene to reveal the main character traits of her protagonist and place him in a troublesome situation to hook the audience into the story. A discussion about the opening of *Midnight Cowboy* helped her to focus on this point.

The early chapters of the novel are devoted to explaining Joe's background. We learn, for example, how Joe Buck became interested in being a cowboy and how his aunt's cowboy boyfriend was the only real father Joe knew. But in the screenplay, it isn't necessary for us to know all this detail about him. The screenwriter has skillfully picked the one important element on which to focus from all this interesting backstory. Without explaining Joe Buck's character, he uses symbolism (the cowboy motif) to represent the protagonist's character flaw. The scene is short; it's condensed; it moves the story along. My student found she could dramatize the conflict between her protagonist and the antagonist by using minor characters in the opening scene with her protagonist. Thus, in one scene, she was able to introduce several characters and the problem of the protagonist. The story had a more interesting opening; it moved along, and it held the audience's interest.

ADAPTING FROM THE SHORT STORY

In this same class, another student came to me. She was adapting a story based on a short story about two identical twin brothers who dislike each other. One twin develops AIDS and the story ends with the mother coming to grips with the fact that one of her sons is gay. The student's problem was that she didn't have enough material to complete her story line. She was in the middle of Act Two and didn't know what crisis to give her protagonist to end the act. She had exhausted the conflict between the mother and son.

In this case, I suggested that the student throw away the original story idea and start over—use the short story as a starting point and retain any elements that might lead to an interesting story line. I thought the conflict between the two brothers could be better used. In her original screenplay, the student followed the short story and focused on the conflict between the mother and son. But why not add this additional conflict into the story even if it wasn't part of the original story? If this addition makes for a more interesting screenplay, then it doesn't matter if we're not being faithful to the original material. Once the student understood this concept, it became easier for her to develop further complications for her protagonist.

Dying of AIDS, abandoned by his mother, and not able to get along with his brother, the character discovers that the only way he can live is if he has a bone marrow transplant from someone with his type of marrow. The identical twin is the ideal candidate. But the brothers hate each other. Here's a chance for character growth as the protagonist finds a way to make peace with his brother, and his brother finds a way to help his dying twin. It could be potentially melodramatic, but if the emotions are treated realistically and the story is structured correctly, it might make an interesting movie-of-the-week. The point is that the original short story was limited. The writer must bring her own imagination into play to work within the framework of a modern-day screenplay.

Another student was adapting "In Another Country," a Hemingway short story dealing with wounded soldiers during World War I. The protagonist of the story is in a hospital with other men who have been wounded in the war. One man, a great fencing master, has been wounded in the hand; another man has a leg wound. New machines in the hospital are being tested on the men and the doctors are confident that these machines will restore the men to their original conditions.

The protagonist, unlike the other soldiers, does not receive his wound due to an act of heroism. It was accidental, and he is scheduled to go back to the front. The Major, the character with the hand wound, is teaching the protagonist Italian. In the climax of the story, he lectures the protagonist about getting married while he's still a soldier. In the end, we learn that the Major has waited—until he is an invalid and out of the war—to be married. The irony is that his wife dies of pneumonia. She was expected to live.

This five-page story has an interesting twist at the end, but the student was having problems with its structure. The problem is that the story is very static. It has interesting characters, an interesting issue involving the war and the doctor's machines, and conflict between the protagonist and the other men who have received their wounds under different circumstances and look down on the protagonist. I suggested that the adapter use the basic kernel of the story, but change some of the plot details. Make the protagonist's problem not his wound, but the fact that he has to go back to the war. This problem is aggravated by the fact that the other men think he's a coward and the protagonist is thinking of getting married.

The protagonist might be looking for ways to prolong his stay in the hospital or get out of the service all together. He might try aggravating his wound. The Major might be aware of this action and be conflicted between his honor and his friendship with the protagonist. It might also be necessary to introduce some additional characters, such as the protagonist's fiancée. This could give us scenes with additional conflict. She might be urging him to get married and he might be resisting, wanting to wait until the end of the war. This character could also intensify the protagonist's internal conflict. I would also change the ending and use the twist about the Major's wife's death. Instead, I would make the protagonist's wife die in the end. Perhaps, instead of trying to be too careful and avoid death, the protagonist finds it coming to him in another form, even though he has gone to all these lengths to play it safe by avoiding the war. The introduction of the fiancée as a character we get to know and like only serves to make her death more poignant than in the short story.

The point here is that the static quality of the short story has to be given an action that will enable the story to follow a linear progression of rising action. If it's necessary to change details of the short story to achieve that aim, then it is a perfectly acceptable practice. Just how far an adaptation can wander from its original source can be best illustrated by looking at Paddy Chayefsky's *Holiday Song*.

Holiday Song

Paddy Chayefsky is known as the writer of *Marty*, *Network*, and *Altered States*. Like most writers he started writing short dramatic pieces for TV. In one such teleplay, *Holiday Song*, we see how radical a departure a dramatization can take from its original source. Chayefsky (1954, p. 35) describes the adaptation process as follows:

> Actually the material to be adapted is rarely of any use to the adapter. It's usually in the shape of a novel or a short story, two literary forms that do not lend themselves comfortably to dramatic exploitation. The convenient introspection of the prose style is difficult to capture in dialogue and action, and the adapter usually winds up picking a kernel of a story from his material and improvising his own characters and insights.

It's interesting that Chayefsky finds the novel and short story's introspection a stumbling block to adaptation. The screenwriter doesn't have the luxury of explaining character motivation and thoughts. She has to use action to show this behavior. She has to wade through the lengthy backstory and pick the necessary information that explains her character, or she has to create her own information if it is lacking in the short story.

As Chayefsky tells us, the original idea for *Holiday Song* came from "It Happened on a Brooklyn Subway," a story in *Reader's Digest* about a photographer who, while riding on a subway in New York, meets a man who was a victim of the Nazi concentration camps. Recalling a similar meeting earlier with a woman, the photographer is able to reunite the separated husband and wife. But Chayefsky, who believed that his adaptation of this piece was basically the same in terms of the work and effort required in writing an original said (Chayefsky, 1954, p. 36):

> The incident is not a good one for dramatic purposes. Despite the fact that it involves bringing together a husband and wife who believed each other dead, there is little emotion involved. There is a certain element of suspense, but suspense is a poor substitute for drama. It would be possible to use this incident as the opening scene of a story and then follow the course of events resulting from

> the meeting of a long-separated husband and wife. But this isn't what the subway incident is about. The incident has only one dramatic meaning and that is: There is a God.

So what is the point of this idea? Chayefsky finds the theme of the story one of its most important elements. This is the kernel Chayefsky talks about; he builds his own version of the story around this theme. But, to prove that there is a God, one has to begin with the protagonist believing that there is no God. Then, when the protagonist reunites the couple and reestablishes his faith, he experiences character growth.

Here is where Chayefsky makes his second major departure from his source material. He changes the occupation of the protagonist. What type of person would be a more effective character for dramatization of a loss in faith than a photographer? There is nothing at stake for a photographer who has lost his faith, unless it affects others in the story. Chayefsky's answer was to change the photographer to a religious figure who has more invested in his belief in God. Such a character might be a minister, priest, or rabbi. Because Chayefsky was Jewish, he wrote about what he knew best and picked a cantor, similar to a rabbi in the Jewish religion. But what would be the driving force pushing the action along through three acts?

Something has to be at stake in the story to maintain interest and suspense. As Chayefsky (1954, pp. 36–37) said:

> What is missing is the story or what I call the urgency. Why is it so urgent that the photographer regain his faith in God? What terrible consequence will occur if he doesn't? . . . All that I want to bring out is how far afield an adaptation wanders. . . . Frequently, the material to be adapted provides no characters at all; or if there have been delineated characters, they are usually impossible to accept. . . . I wanted to tell a charming folk tale about a small Jewish community struck by the catastrophe of its cantor refusing to sing for the High Holidays.

So the urgency in this story becomes the High Holidays and the faith of the entire congregation. If the cantor loses his faith and refuses to sing, then all the members of the synagogue may lose their faith as well.

In a 30-page or half-hour adaptation, as in *Holiday Song*, it's important for the writer to begin his story as soon as possible. The story begins by establishing urgency with a camera shot of an announcement that the cantor will be singing at the upcoming High Holiday services. The next scene quickly establishes the cantor, his surroundings, and the fact that something is wrong. Remember our structure of the troublesome beginning that escalates into the Act One problem break. Here, we find the cantor on his bed bemoaning his loss of faith in God. A friend, Zucker, comes to visit the cantor and implores him to seek spiritual help from a famous rabbi in New York City. The cantor's loss of faith deepens when he gets on the wrong subway and meets a man from Hungary who was in a concentration camp and lost his wife and children. No God could allow such horrible things to happen. There must be no God.

The cantor's problem has worsened at the end of Act One. Note how the short story version only gives us this one main scene. In the adaptation, we add our own initial scenes and build to the scene in the subway. In Act Two, the cantor attempts to see the rabbi one more time. He asks directions to the train from an attendant on the platform. He is once again directed to the wrong subway car where he meets a woman who also is a concentration camp victim. Once again, the cantor loses his faith and faces a spiritual crisis at the end of Act Two. Note the scene between the cantor and the wife, only alluded to in the short story, is added into the screenplay.

In Act Three, the cantor, during the climax of the story, succeeds in reuniting this lost couple. In the resolution scene, he goes back to thank the attendant who placed him on the wrong train two times. There is no such person. The attendant, the cantor concludes, was God. The cantor experiences character growth and in the final scene, also added to the script, we see him singing in the synagogue on the High Holidays. Although the script is simplistic by today's standards, it does illustrate the point that the adapter of a short story must be prepared to add her own material, characters, and scenes in order to fit the dramatic structure so simply illustrated by this early TV drama.

2001

Another story that illustrates just how far afield a screenplay can wander from its original short story is Stanley Kubrick's *2001*. Based on Arthur C. Clarke's short story, "The Sentinel," Kubrick takes the simple premise of the short story and expands it into an epic story about space travel and visitors to our galaxy.

Basically, the nine-page short story deals with a space exploration to the moon. The men on this journey find an object they cannot explain. It has been made by some advanced intelligent society that visited our galaxy before people evolved. The purpose of this object was to signal this advanced race that another intelligent life force was evolving in the universe. In Agel (1970, p. 22), Clarke says:

> Perhaps you understand now why that crystal pyramid was set upon the Moon instead of on the Earth. Its builders were not concerned with races still struggling up from savagery. They would be inter-

> ested in our civilization only if we proved our fitness to survive—by crossing space and so escaping from the Earth, our cradle. That is the challenge that all intelligent races must meet sooner or later.

The object becomes a sentinel or beacon to other races and informs them of the development of a new intelligence in the galaxy.

In Kubrick's adaptation of this story, the initial discovery of the sentinel is retained and becomes one of the early scenes. What is interesting about this adaptation is how Kubrick, like Chayefsky, used the story as a jumping-off place for the main direction of his screenplay. In Kubrick's case, the entire movie is constructed from the basic premise that an intelligent life force has set up this beacon to contact earth. The remaining seven-eighths of the movie is then devoted to trying to contact the intelligent source of this object.

The men traveling to the space station, the use of the computer HAL, and the climactic series of scenes when the men reach Jupiter have been totally contrived by Kubrick based on his concept of what future space travel would be like. In Agel (1970) Clarke said, "I would say that *2001* reflects about ninety percent on the imagination of Kubrick, about five percent on the genius of the special effects people, and perhaps five percent on my contribution" (p. 136).

Interestingly enough, the early reviews of *2001* were far from flattering. Renatta Adler of *The New York Times* found the movie to be a complicated, languid movie "in which almost a half-hour passes before the first man appears and the first word is spoken, and an entire hour goes by before the plot even begins to declare itself" (Agel, 1970, p. 207).

Stanley Kauffmann of *The New Republic* found the problem with the movie to be its handling of the adaptation (Agel, 1970, pp. 223–24):

> Part of the trouble is sheer distention. A short story by Arthur C. Clarke has been amplified and padded to make it bear the weight of this three-hour film. . . . Kubrick had to fill in this lengthy trip with some sort of action, so he devised a conflict between the two men and the giant computer on the ship. It is not exactly fresh science fiction to endow a machine with a personality and voice, but Kubrick wrings the last drop out of this conflict because something has to happen on this voyage. . . . It states one of Clarke's favorite themes—that, compared with life elsewhere, man is only a child; but this theme, presumably the point of the whole long picture, is sloughed off.

Although *2001* has emerged from these early negative reviews as a classic example of filmmaking, it does illustrate two key points. In the first place, it is essential that the adapter of a short story add his own material into the story to make it work. Perhaps it is the conflict between this extra material and the intent of the original writer that makes the movie fail in certain key ways. In some ways, the critics are correct that the film is overindulgent. Kubrick doesn't exactly follow dramatic structure as we have talked about it up to this point in this book. The slow beginning and unnecessary scenes do tend to make the movie drag on more than it should.

Like many contemporary films, there is a tendency to rely on special effects as a substitute for story line. While the movie will remain a great one, one can only speculate on what a tighter story line and more adherence to the Clarke theme of the man–child concept might have done for the movie. The second key point, however, is the fact that the movie is able to make so much out of the brief short story source. This does attest to the genius of the adapter.

"Rita Hayworth and the Shawshank Redemption"

Stephen King's short story is a long one, numbering over 100 pages. In a sense, it is almost a short novella, which presents problems common to both the short story and the novel. In general, the screenwriter is both forced to add situations, characters, and scenes as in a short story and condense backstory, characters, and scenes as in a novel.

Like the novel, large portions of backstory are omitted in the adaptation. As mentioned previously, backstory on Red as well as Andy is omitted. Details about Andy's trial are omitted—"A clerk from the Wise pawnshop in Lewiston testified that he had sold a six shot, 38 police special to Andrew Dufesne just two days before the double murder. A bartender from the country club testified that Andy had come in around seven" (King, 1995, p. 19). In addition, details of the investigation into the murder are cut and scenes of Andy drinking following the revelation of his wife's infidelity are cut. Some of the elements of prison life are also cut—the information about Andy's drinking and the selling of dope and pills while there.

On the other hand, certain elements are added from the short story to increase its dramatic value. The subplot involving Andy's friend from the library, Brooks, has been changed in the adaptation process. In the novel, Brooks leaves prison and the story ends. In the movie, Brooks kills himself on the outside. His failure to adapt to his return to the real world enables the screenwriter to develop his theme of institutional life and provides additional motivation for Andy's escape. Interestingly enough, it also adds tension to the character of Red. When he is released from prison, he is put in the same room and job as Brooks (another added element in the movie). Will he also be a victim of institutional life or will he succeed in finding Andy in Mexico? The writer uses the Brooks' incident to add suspense to the conclusion of the movie. The final scene, also added to the movie, answers these questions as we see Red strolling down the beach in Mexico while Andy works on his boat.

SUMMARY

Unlike the adapting of a novel into a screenplay, the adapter of a short story has to be more prepared to call on his imagination. In most cases, as Chayefsky stated, the process of adaptation requires the same mistakes, struggles, and imaginative processes as the writing of an original. Unless the adapter is willing to call on imagination when rewriting the material, his efforts are probably doomed to failure.

Whether one writes an original or adapts from a novel or short story, or from the stage, the problems of creating workable drama are both a challenge and an adventure. Trying to duplicate the work of the original piece only sells short your own creative potential. We all want to remain faithful to the work of another artist, but as artists ourselves we are involved in creating our own works. The adaptation of the short story isn't meant to be a duplication of the original story. It's meant to be the springboard for the creation of a new piece of cinematic art.

EXERCISES

1. Find several short stories that have possible dramatic potential. Discuss in class how these stories might be expanded into a screenplay, the new characters you might add, and the problems the short story presents for the adapter.
2. Reread Clarke's "The Sentinel." Can you suggest a story line that takes a different direction than the one used by Kubrick? Is your story tighter? Have you devised any new conflicts or a more up-to-date issue than a computer gone crazy? Discuss these ideas in class. Can you devise a faster beginning for the screenplay that would hook a contemporary audience more quickly?
3. Can you find any other short stories that have been adapted into screenplays? What are the differences between the original and the screenplay adaptation? Discuss these differences and the reasons for them in class.
4. Read D. H. Lawrence's short story "The Rocking-Horse Winner" about a young boy who is able to pick racing horse winners by riding on his toy rocking horse. Compare the structure and ending of this story with the movie adaptation. What changes were made and why? How would you adapt this piece? What elements would be most useful in a dramatization of a story of this type?

9

Plays

Up to this point we have been looking at the problems involved in making the transition from the novel to the screenplay. Sometimes the source material for a screenplay is a theatrical play. Sometimes the adapter is the writer of the original play. What problems are involved when starting with a play? Is it easier to adapt from a play, since a play follows many of the same dramatic rules as the screenplay? Are the similarities between the play and the screenplay greater than the differences between the screenplay and the novel? Which plays have been successfully adapted into screenplays? This chapter examines some of these questions and others that may arise as one adapts a play into a screenplay.

We will begin by examining some plays that have been adapted into successful screenplays and looking at the various steps involved in the process. Some guided exercises will give you practice in some of these adapting techniques. We'll conclude by looking at one film that was ultimately adapted into a play. Sometimes, the adapter is the playwright. (This is more likely to occur with a playwright than with a novelist because the two genres are so different.) Sometimes the adapter is a new writer. In any case, the difficulties of adapting one's work or another writer's work are similar and can extend to the writer of the original screenplay or theatrical play as well.

STAGE VERSUS FILM

Many of the dramatic elements discussed so far in this text can also be found in a play. A play, for example, needs a sympathetic protagonist, conflict, problems for the protagonist, an interesting subplot, and an important issue to deal with. On the other hand, because of the physical structure of a theater, there is a degree of compactness in a play that is missing from a screenplay. Most of the scenes in a play, for example, need to be located in one or two places to avoid elaborate set changes. This arrangement creates more tension and conflict between characters, because the characters, unlike in film, literally cannot escape each other.

Characterization is also different in a play than in a screenplay. As with a novel, the writer has more time to develop characters and spend more time on backstory. Dialogue can also be different. Speeches are often longer in a play; characters are more likely to talk to the audience or deliver soliloquies. Structurally, there are fewer scenes, sometimes different endings, and even different issues involved in the story. Technically, there are differences as well. In film, there are camera angles and editing. On the stage, there are elaborate spectacles, interesting costuming, and sometimes even live music.

On the stage, the audience is basically watching the performance from one angle: a medium or long shot from their seats. Here, the importance of dialogue and the actor's verbal ability are the keys to a good performance. In film, the audience can watch the performance from several angles. They can see a close-up of the protagonist's face and watch reaction shots of the other characters in the story. Dialogue can play a secondary role, with key communication coming through the visual—the specific reaction shot, a zooming in, or a close-up of the character's eyes.

Adapting Shakespeare

The adapter of a play needs to be aware of the advantages and disadvantages the technical aspects of cinema can add or detract from his work. Nowhere is this more apparent than in some of the attempts to adapt Shakespeare to the screen. In recent years several new adaptations of Shakespeare have appeared, such as the recent four-hour rendition of "Hamlet" and two unusual versions of "Romeo and Juliet."

The greatest problem with a Shakespearean adaptation is poetic language. Some adapters have attempted to deal with the language by updating it to make it more accessible to a mass audience. For the most part, these adaptations have failed. For those who are familiar with Shakespeare, such attempts totally distort the essence of the dramatist's original intention. One older adaptation of "Hamlet," for example, tried to prune and modernize Shakespeare's work by omitting subplots and characters. The well-known minor characters of Rosencrantz and Guildenstern were cut from the movie version, giving us a film that was very different from the Shakespearean "Hamlet" to which most people are accustomed.

Kenneth Branagh's 1996 adaptation of "Hamlet" remains faithful to the original but suffers from excess length and reliance on movie stars for key roles. In the recent adaptation of "Romeo and Juliet," the writer attempts to update the story by making it take place in a modern city, Los Angeles, and turns the feud between two families into a battle.

If one is going to adapt a Shakespearean play, or any play for that matter, mixing the dialogue with the visual is often a problem. Many adaptations have made the mistake of cross-cutting the dialogue with over-the-shoulder shots and close-ups. Sometimes, especially for a Shakespearean play in which understanding the language is difficult, this makes the dialogue choppy and unnatural. Sometimes the camera is too obtrusive and duplicates what the character is saying in dialogue. "Look at that knife," with a close-up on the knife is redundant. The key to successful adaptation of theatrical dialogue is to use the camera to visually give us the same information without duplicating both the visual and the content of the dialogue.

There have been more than one adaptation of some Shakespearean plays. "King Lear" is one of the most accessible of Shakespeare's plays and has been adapted for a film several times. A related or more modern attempt is the film *The Dresser*, which deals with a modern actor playing Lear on stage. In the more literal adaptations of "Lear," we can see how the adapter or director has successfully used the advantage of the camera to symbolize or reveal character information about Lear.

In a 1970 film version of "Lear," the famous storm scene is shot from a plane, with Lear seen as a small figure on the heath dwarfed by nature. In a 1971 version of the same play, the storm scene is shot with a close-up of Lear's face superimposed over the sounds of the storm raging in the background. The message is that the storm symbolizes the craziness raging inside Lear's head. In the first version, the conflict was between Lear and nature. In the second version, Lear's conflict and the problem that caused alienation with his daughters were within himself. The camera gives us two very different but rich interpretations of the same scene.

The writer can use a camera angle to give us information about the character without distorting the playwright's original intention. In 1997, another unusual adaptation of "Lear" appears in the movie *A Thousand Acres*. Here, the basic story of a father and his three daughters is placed into a modern setting on a country farm with Jessica Lange and Michelle Pfeiffer playing the ungrateful daughters. Jason Robards plays the mad "Lear" character as a farmer who gives his farm to his three daughters.

In the adaptation of "Rear Window," the writer uses narration at the opening of the short story to explain how the protagonist broke his leg. In the film, the camera pans the room and shows us the protagonist's broken leg juxtaposed to a shot of a news story. (Note the protagonist is a reporter and broke his leg on an assignment.) No dialogue is ever used. Dialogue can slow a film down, but a visual shot can move the story along.

Other problems in filming Shakespeare, or other classical plays for that matter, are their commercial accessibility. Some may argue that Shakespeare is not for the general public, but the truth of the matter is that none of the adaptations have really made much money. Although, with the changing marketplace and demand for more cable material, adaptations of classical plays probably will find a more specialized audience. But what about the problem of the poetic language? Should one adapt the poetic language as written and wind up with a filmed version of a stage production, or should one attempt to update the language and distort the original source? This is not an easy question and is more difficult to answer than with a novel or short story adaptation. Most adapters have taken the middle road and try to keep most of the original language without just filming the stage production, making changes in language to accommodate the camera and meaning of the piece.

Another problem with this type of play is the many theatrical conventions. How, for example, does one render a soliloquy or characters who break the fourth wall by talking to the audience? Soliloquies are difficult to bring off, because film is really about the clash and exchange of dialogue. Although characters in films can make longer speeches similar to soliloquies, these usually occur later in the film when the character has an emotional, revealing moment. This is often done under the guise of a conversation, such as in *Kramer versus Kramer* when Ted is on the witness stand and makes a long speech about the rights of fathers. This speech comes in response to a question from Ted's attorney and appears to be part of his conversation with the attorney rather than a long soliloquy. In terms of breaking the *fourth wall* (when a character talks directly to the audience), there are many examples of films that use this theatrical device. We find it used in *Head over Heels*, *Ferris Bueller's Day Off*, and Woody Allen films. It serves mainly as a comic relief rather than as a means for a character to reveal inner emotions.

Adapting Other Plays

As a writer moves away from classical to more modern theater, many of the problems of adaptation are similar to the problems one encounters when adapting from the short story or novel. Consideration of the audience is always a problem in any adaptation.

"Marvin's Room"

The adaptation of "Marvin's Room" eliminates large amounts of backstory about the various characters. We lose information on Lee, her schooling in cosmetology, and her relationship with her dead husband Hank. Her move into the convent and information about the nuns and Ruth are also cut. We lose the backstory about Ruth's health problems, her relationship with God, and her philosophy on why God made her sick. Backstory on Hank is also cut. His life in the institution is touched on in the movie but in the play it is explored in more depth (McPherson, 1992, p. 60):

> Most of the time I keep to myself. Most of the time I sit in my room. I've got a roommate, but most of the time he's got his face to the wall. Most of the time I think about not being there. I think what it would be like to be someone else. Someone I see on the TV or in a magazine, or even walking free on the grounds. They can keep me as long as they want. It's not like a prison term. I've already been there longer than most.

Adaptations of plays, like novels, need to be made visual. Thus, information on Hank in the institution is shown in scenes of Lee visiting him in the hospital. We see him alone in his room rather than hearing him tell us about it in the play. Movies need to "open up" plays, which have only a few locations; thus, scenes are added to movies to accomplish this purpose. We see Hank and Lee arguing in the car rather than hear about it. We see Hank running away from the house rather than being told about it.

"A Streetcar Named Desire"

In the adaptation of Tennessee Williams's "A Streetcar Named Desire," we find that many of the homosexual references of the play have been eliminated for the film version so that it is more acceptable to the mass audience. We also find that the ending has been changed to make it more acceptable, with most of the characters turning against Stanley in order to see him punished for his macho attitude toward women. Even Stanley's character is rounded out to make him more sympathetic to the audience.

"Who's Afraid of Virginia Woolf?"

In "Who's Afraid of Virginia Woolf?," much of the explicit sexual language has been cut out of the adapted version to make it less offensive to the movie audience. Like many adaptations of stage plays, scenes are added to open it up to the visual possibilities of the cinema. The scenes with George and Martha, which take place in the living room in the play, take place outside the house in the film.

"Front Page"

In the film adaptation of Charles MacArthur's "Front Page," the protagonist, Hildy Johnson, was changed from a man to a woman. This was done because of the play's homosexual overtones regarding the love–hate relationship between Hildy and his boss. By making Hildy a woman, the writer eliminates this problem and adds a love interest, which always holds more appeal for a film audience.

Many film productions are basically duplications of the original staged productions from Shakespeare to *On Golden Pond* to *Educating Rita* to Pinter's *Homecoming.* Those that remind you of the original and those that totally ignore the original seem to be the most common. In my opinion, the adapter needs to walk a very straight line when dealing with adaptations and aim, even more so than for the novel or short story adaptation, for something in the middle of these two extremes. Nowhere is this phenomenon more apparent than in the works of Neil Simon, perhaps the definitive and most commercial American playwright–screenwriter of the 1970s and 1980s.

"Biloxi Blues"

Nothing illustrates the problem of walking a middle line and the compactness of the play versus the expansiveness of the screenplay than Neil Simon's "Biloxi Blues." Basically, the play and screenplay open with the same scene—the main characters are on a train heading to Biloxi, Mississippi. Of course, in the Simon play (1988), the scene is not true to reality: "All set pieces are representational, stylized and free-flowing" (p. 3).

Following this disclaimer, the play goes on to introduce all the main characters. The major difference between the play and screenplay lies in the description of some of the characters. The description of Roy, for example, who smelled like a tuna fish sandwich, is omitted from the screenplay, along with Carney's singing of "Chattanooga Choo Choo" and the comparison of Carney to Perry Como. As in the novel, the playwright has a tendency to describe his characters in more detail than the screenwriter. In the play, so much of the story development comes through characterization that the writer is apt to spend more time delineating his characters. In the screenplay, however, the descriptions are briefer, broader, and allow for many different actors to play the various parts.

Structure

In the play, the transition from the first scene on the train to the second scene in the barracks is done with Carney's singing. In the screenplay, the next scene occurs with the men arriving at the base. Because this is a movie, the writer takes advantage of the visual element of film by taking the scene outdoors. This change fits in nicely with Jerome's line: "This is like Africa hot. Tarzan couldn't take this kind of hot" (Simon, 1988, p. 9).

The line is more effective in the movie because we can actually see how hot it is in Mississippi.

Dialogue

In the next scene, we meet the character of Toomey, the eccentric drill sergeant. The speech he gives to the men is basically the same as the speech that winds up in the screenplay, except that certain sections have been omitted. In the play, Toomey tells the men, "There's only one way to come out of a war with a healthy body and sane mind, and that way is to be born the favorite daughter of the President of the United States" (p. 11). Later, when describing his injury in the war, Toomey says, "This injury has caused me to become a smart, compassionate, understanding and sympathetic teacher of raw, young men or the cruelest, craziest, most sadistic goddamn son of a bitch you ever saw . . ." (Simon, 1988, p. 11).

Note that the tendency in adaptation, whether it's a play or novel, is to condense. Here as with a novel, the backstory on Toomey is condensed for its inclusion in the screenplay. Long pieces of dialogue that work in a play need to be cut to work in the screenplay, which requires a faster pace to hold the audience's attention. Cutting dialogue also eliminates the problem of static stage scenes.

In this same scene, the exchanges between Toomey and Eugene are condensed. When Toomey questions Eugene in the play, he says, "In my twelve years in the army, I never met one goddamn dog face who came from 1427 Pulaski Ave. Why is that Jerome?" Eugene answers, "Because it's my home, only my family lives there. I'm sorry, I meant I live in Brighton Beach, Brooklyn, New York" (p. 11). Again, this exchange is unnecessary. It's interesting backstory, but it only slows down the action of the screenplay. Backstory on Wykowski being a furniture mover is also omitted, as well as Toomey's reply, "That's just what they need in the South Pacific, Wykowski. Someone who knows how to move furniture around in the jungle" (Simon, 1988, p. 11).

Further backstory on Carney and the shoe store is also omitted. One interesting change from play to screenplay involves the switching of Private Carney to Eugene, as the man Toomey singles out to pick the men to do punishment. This serves to immediately intensify Eugene's, the protagonist, problem.

Character Action

Even when the antagonist, Toomey, focuses on Jerome as the cause of the platoons' problem, the approach to presenting this situation differs from play to screenplay. In the play, Toomey uses dialogue to single Jerome out: "As the sweat pours off your brows and your puny muscles strain to lift your flabby, chubby, jellied bodies, think of Private Jerome of Brighton Beach. . . . Fate always chooses someone to get a free ride" (Simon, 1988, p. 19). In the screenplay, Toomey's long speech is omitted. The screenplay uses action and visual moments to stress the growing conflict within the platoon and their hatred of Jerome.

The next scene is the same for both screenplay and play, with the men being introduced to Army food. The major difference is that the scene in the screenplay is considerably shorter than the one in the play. As with previous scenes, there is slight character modification, which in this scene places the emphasis of the problem more on Eugene and less on Arnold. In the play, for example, Epstein, not Eugene, is forced to eat everyone's food; in the screenplay, Eugene is forced to eat the disgusting dinner. The moment of Epstein screaming, "I won't eat slop. I won't eat slop" (p. 27) is omitted from the screenplay. The intention is to redirect the conflict away from the different members of the platoon, who collectively are the protagonists of the play, to one major protagonist, Eugene, in the screenplay. Not only does it help to focus the audience's attention on the issue of the piece, but it also makes the story move at a quicker pace.

This change in emphasis from Arnold to Eugene can also be seen in another scene omitted from the play. In this case, a group of soldiers come into the latrine while Arnold is cleaning. After a brief confrontation, they throw him upside down into the toilet. Again, Arnold is not the main protagonist of the screenplay; therefore, this scene is omitted to redirect the problem to the main protagonist, Eugene.

Like a novel, a play has the advantage of using narration. In the play, the scene involving the men marching into the Mississippi swamp at night is told through narration. As Eugene says, "The only time I heard strange sounds like that was at Ebbets Field when the Dodgers played. Toomey, make Wykowski carry me the whole fifteen miles" (Simon, 1988, p. 27). In the screenplay, the writer doesn't talk about an event, he can show it by taking advantage of the visual element of film. Interestingly enough, the screenplay's scene has a certain unity. In the first place, by making the scene take place during the day rather than at night, the screenwriter once again establishes the oppressive heat and uncomfortableness of the Mississippi backdrop that we saw earlier when the men arrived at the training camp. In addition, by having Jerome whisper into Toomey's ear regarding Wykowski's push-ups, the writer is able to harken back to an earlier scene of the screenplay in which Toomey has Jerome pick another man for push-ups. This change intensifies the protagonist's problem and visually and structurally causes the action of the screenplay to continue to rise. **[Exercises 1–4]**

Condensing Dialogue

The chief way to condense scenes in the transformation from play to screenplay is the omission of key speeches. Often the writer will cut down longer individual speeches of some of the characters but keep relevant parts. This is true in "Biloxi Blues" in which two key speeches by Eugene and Toomey are typical of the writer's approach to dialogue in his adaptation. In the first speech involving the bet between the men in the platoon, Eugene asks how each man would spend the last week of his life if he knew he was about to die. Eugene's introduction to his fantasy is cut short in the screenplay, which omits the section about wiping out a Japanese battalion, getting a statue in Brighton Beach, and a junior high school named after him.

In the next scene, Toomey confronts the men about the supposed theft of one of the men's wallets. In Simon's play (1988, p. 43), Toomey begins:

> I have put up with everything from mutiny to sodomy . . . sodomy is the result of doing something you don't want to do with someone you don't want to do it with because of no access to do what you want to do with someone you can't get to do it.

He then goes on to talk about the theft and concludes by asking the guilty party to confess. Note, when adapting his own work, Simon shortens Toomey's amusing speech, choosing to begin with the next key element that moves the story along, namely Toomey asking the guilty party to confess.

Toomey confronts Epstein for confessing to stealing the wallet in the following scene: "How the hell do you think you can beat me?" (p. 47). Epstein replies that he doesn't want to fight the sergeant, but to help him with the men. Toomey finds this amusing; he thinks Epstein is low on smarts. Epstein replies, "I don't think it's necessary to dehumanize a man to get him to perform" (Simon, 1988, p. 47). Interestingly enough, this scene is also shortened. The writer goes from Toomey's question to Arnold's crucial response about dehumanizing the men. The interesting repartee between the two men is cut to make the scene move more quickly. **[Exercises 5–7]**

"Biloxi Blues," Act Two

As in Act One of "Biloxi Blues," Act Two follows the same adapting principles that we have discussed so far. The screenplay continues to open up the scenes from the play. We see the men in the town of Biloxi on their way to the whorehouse, for example, rather than opening directly with the whorehouse, as in the play. The scene with Hennesey being arrested is also opened up. In the play, the arrest occurs in the barracks. In the movie, the men are hiking in the Mississippi heat when the MPs suddenly drive up and stop the march and take Hennesey away. The scene is more dramatic this way and underlies the theme of the suddenness of life and death, underscored by Eugene's bet in Act One. At any time, something can happen to change a man's life. That's part of the reason it's so important for Eugene to have his sexual encounter and fall in love during Act Two. During wartime, life is short.

Backstory

Act Two also follows Act One in elimination of extensive backstory. Thus, in the screenplay, we lose the backstory about Carney and his girlfriend in Albany. We lose the information about Daisy's family really being from Chicago and how her father got a job with a newspaper in New Orleans. This job eventually got them relocated to Biloxi. This information adds depth to Daisy's character in the play, but is unnecessary in the screenplay for the writer's exploration of Eugene's problem, which isn't his relationship with Daisy but his conflict with Toomey. Toomey's motivation for being a strong disciplinarian (he was spanked by his father) is also omitted from the screenplay. Again, it isn't necessary for us to understand Toomey's actions; it's more important for Toomey to be eccentric and to act with no logical reason in the screenplay. His behavior reinforces the theme about the delicate balance of life during war, when events can happen for no explainable reason.

Endings

The final difference between the play and screenplay is in the ending. In the play, Toomey confronts Epstein and threatens to blow his brains out. In the screenplay, Eugene replaces Epstein in this scene. In the screenplay, it's important for the protagonist to confront the antagonist in the climax scene, and, of course, ultimately solve his problem. Eugene does stand up to Toomey and experiences character growth when he realizes how much he misses Toomey after Toomey is transferred to the hospital. Eugene also grows when he realizes how his years in the Army were some of the best of his life.

In the play, we find out how each man survived the war, but the screenplay ends more ironically. The men never make it to the war. The war ends before they finish basic training. Instead of getting their war history, the screenplay concludes with a description of what happened to each man later in his life. Eugene becomes a writer, Epstein becomes a lawyer, Wykowski develops stomach problems, and Carney becomes a teacher. Epstein becomes the most feared man in New York, especially by the Mob. **[Exercises 8–9]**

OPENING UP

A classic term applied to the process of adapting a stage play or novel into a screenplay is *opening up*. The reversal of this process is termed a *closing down*, whereby the flighty visuals of the cinema are encapsulated into the tight restraints of the theater. The beginning writer can be misled by this terminology or rely too heavily on it when adapting from play to screenplay or from a screenplay into a play. The writer may confuse the external space she associates with cinema with the process of expansion, and internal theater space with the process of contraction. Although it's true in the examples cited earlier in this book and in this chapter that most writers seem to open up from the play to the screenplay, this is not always the case. One must learn the rules, first, before tackling the exceptions.

The adaptation for the film *Amadeus* did not open up from the play to show us the streets and environs of Vienna. Instead, we are shown the inner spaces of the palaces, music rooms, and theaters of the city in grand detail, because this is the space in which Mozart lived and worked. *La Cage aux Folles*, a screenplay adapted into a play, was much more open to different locations than the play. In the film, we see the main characters in the seaside community and the bride's parents in their Gothic, formal home. This use of space adds contrast to the character's lives and their real attitudes. The play, however, could not easily take us to these different locales, so character projection, costume, and music were relied on to show these contrasts.

FORM

The first thing a writer should consider when beginning to think about the structure of his work is the story itself. A writer should never become so dominated by form that he forgets the basic story line of the material. Stories, for the most part, can be told in either a continuous or an elliptical way. Films are circular in nature, told in a long, winding curve that ends and begins on the same lonely crosswalk on some isolated country road. Films generate unity. Stage, on the other hand, is often continuous in its sense of the present, as characters move from moment to moment without the quick cuts provided by film editing. The "Amadeus" stage play is a story told as a remembrance or confession. In film terms, this clearly is a flashback, and the translation held no surprises for the adapter. Once the narrator is in past time, the memories have logical literary unity; for the most part the film retains this unity.

The movie version of *La Cage aux Folles* made only occasional trips to Paris where the developing conflict with the sanctimonious, political, self-serving father-in-law heightened the comedy and suspense. In the theatrical adaptation, the plot points provided by the future father-in-law are less farcically portrayed and more evenly integrated into the dramatic piece. In some ways, this change is an improvement over the original in that it leans more toward the real than the farcical, a choice that supports the underlying message of the story. In the real world, differences of sexual preference do not deny a common humanity, and one has the ability to love in, at least, the universal sense. Thus, the type of plot one is dealing with, plus the importance of plot in supporting the message of the story, becomes the ultimate consideration in choosing format and technique. One might opt for a continuous stage structure in a screenplay if that type of structure works for the material. The choice of format does not exclude the techniques one might employ in the genre from which one is adapting.

CHARACTERIZATION

Characterization is one of the prime considerations in the adapting process. This is true in adapting from a play to a screenplay. On stage, the actor projects the characterization, literally throwing it out to the audience, constantly providing emphasis and directing the audience's attention to particular facets of that character. In film, however, the audience's attention is literally focused by the writer and by the camera. Whereas the stage actor conveys character by projection, on film it is the camera that projects. It is the film actor's job to reveal, but the camera carries much of the dramatic burden through image size and placement. In *Amadeus*, the close-ups of the old man remembering show the swollen blue veins; the brown, worn-down teeth; and other vivid, detailed physical manifestations of not just age, but the damage a corrupted soul renders to the human body.

In *La Cage aux Folles*, close-ups greatly contributed to the believability and affection the audience experienced toward the main character. Although the character is bizarre and outrageous, we see that his tears are real and we delight in the mental tricks reflected in the subtle gestures the camera shows us that are not experienced by the other characters. Note that a good writer supplies the actions of the character in the scene, but is aware of the camera's ability to bring out these actions. The writer doesn't have to direct the scene but should have an awareness of camera movement and language.

In the stage version of "La Cage aux Folles," much of the subtlety of gesture is lost, but it is compensated for by the emotional suggestion and atmosphere provided by the play's score and the sentiment of the lyrics. Again, the trick is to remember what characteristics the story line suggests for your character and then use the technique of the genre you are employing to bring them out to their fullest.

SYMBOLISM

Another major element in the adaptation process is the message of the material with which you are dealing. Symbolism is an important element in both film and stage because it reinforces messages. In *Amadeus*, the black cape costume and mask do more than just conceal the identity of the narrator as he manipulates Mozart. The very presence of this image expresses the scale and scope of the darkness and maleficence of the narrator's soul. The cape scars the man who wears it and its heaviness contrasts nicely to the pastel, ruffled costume of Mozart. This powerful use of symbolic element was first used in the original stage version. The fact that the adapter chose to retain this element in the film, when film has a tendency to use symbolism in a more minor way than theater, speaks highly of the writing skills of the adapter. In the final analysis, the writer must look to the meaning of the material she is working with before making her adapting choices. While it is true that stage lends itself more successfully to symbolism and abstraction of meaning, when these devices are used correctly in film, they can be very powerful. **[Exercises 10–11]**

SUMMARY

The fundamental differences between the stage play and the screenplay are mainly in technique. Depending on the type of material one is adapting, the genre might become a consideration. Plot, character, and theme are definitely elements in that choice. In recent times, more hybrids of both the stage and the screenplay formats have materialized. The appearance of hybrids suggests that good material might ultimately transcend form. Surely the truly creative writer can tell a successful story in any form. For all the analyses of both genres, one should never forget that modern theater has been greatly influenced by modern film techniques, but it is equally true that good narrative film has its roots in the theater.

EXERCISES

1. "Chapter Two," a Neil Simon play (1980), deals with the relationship between a widower, George, and an actress, Jennie. As in most adapted plays, the scenes used in the movie are condensed from the same scene in the play. Take the following scene (pp. 660–61) from the play, which involves a telephone conversation between George and Jennie, and condense it. After you finish, check the appendix to see how close to the screenplay version you've come.

```
GEORGE: No. This is what they call amusing telephone
conversation under duress. . . . So what is it you do?
JENNIE: I'm an actress. (He doesn't respond.) No "aha"?
GEORGE: Leo didn't tell me you were an actress.
JENNIE: I'm sorry. Wrong career?
GEORGE: No. No. Actresses can be, uh, very nice.
JENNIE: Well, that's an overstatement but I appreciate your
open-mindedness.
GEORGE: Wait a minute, I'm now extricating my mouth from my
foot. . . . There, that's better. So you're an actress and
I'm a writer. I'm also a widower.
JENNIE: Yes. Faye told me.
GEORGE: Faye?
JENNIE: Faye Medwick. She's the one pushing from my side.
GEORGE: Leo is getting up a brochure on me. We'll send you
one when they come in. . . . I understand you're recently
divorced?
JENNIE: Yes. . . . How deeply do you intend going into this?
GEORGE: Sorry. Occupational hazard, I pry incessantly.
JENNIE: That's okay, I scrutinize.
GEORGE: Well, prying is second cousin to scrutiny.
JENNIE: Wouldn't you know it? It turns out we're related.
GEORGE: I don't know if you've noticed but we also talk in
the same rhythm.
JENNIE: Hmmmm.
GEORGE: Hmmmm. What is hmmmm?
JENNIE: It's second cousin to aha! . . . You're a very
interesting telephone person, Mr. Schneider. However, I have
literally walked in the door, and I haven't eaten since
breakfast. It was really nice talking to you. Goodbye.
GEORGE: Listen, uh, can I be practical for a second?
JENNIE: For a second? Yes.
GEORGE: They're not going to let up, you know.
JENNIE: Who?
GEORGE: The pushers. Leo and Faye. They will persist and
push and prod and leave telephone numbers under books until
eventually we have that inevitable date.
```

JENNIE: Nothing is inevitable. Dates are man-made.
GEORGE: Whatever. . . . The point is, I assume you have an active career. I'm a very busy man who needs quiet and few distractions. So let me propose, in the interest of moving on with our lives, that we get this meeting over with just as soon as possible.
JENNIE: Surely you jest.
GEORGE: I'm not asking for a blind date. Blind dates are the nation's third leading cause of skin rash. . . . What if we were to meet for five minutes?

2. Read several other plays by Neil Simon. Compare a scene from each play to the same scene in the screenplay or movie version. Can you tell where the scene has been condensed? Why were the lines omitted?

3. In these same Simon plays, can you find instances of elimination of backstory, dialogue, and the introduction of additional scenes to open up the story and make it more visual? Why were bits of backstory omitted?

4. Compare and contrast Simon's other two plays in his trilogy on his early life—"Brighton Beach Memoirs" and "Broadway Bound."

5. In the following "Biloxi Blues" scene (Simon, 1988, p. 49) between Epstein and Eugene, Eugene compliments Arnold for standing up to Toomey. Arnold explains his philosophy of principles. In the screenplay, the scene is considerably shortened. Can you pick the lines that were eliminated and the crucial ones that were kept for the screenplay?

EUGENE: Why do you always have to do things the hard way?
EPSTEIN: It makes things more interesting.
EUGENE: It also makes a lot of problems.
EPSTEIN: Without problems the day would be over at eleven o'clock in the morning. . . . Principles are okay. But sometimes they get in the way of reason.
EUGENE: Then how do you know which one is the right one?
EPSTEIN: You have to get involved. You don't get involved enough, Eugene.

6. Study the confrontation scene between Willie and Al in Neil Simon's "The Sunshine Boys" in which the characters argue over the use of the word "enter." See if you can decide how the dialogue for the screenplay was cut in this same scene. Why did Simon choose to eliminate key speeches from his play?

7. Compare a Shakespearean adaptation, such as "Othello" or "Macbeth," to the original. Discuss the different problems in adapting a classical work to a more modern one. How do special effects enter into the filmed version of *Macbeth*?

8. Compare adaptations of Tennessee Williams's plays "The Glass Menagerie," "Sweet Bird of Youth," and "A Streetcar Named Desire" to

their film versions. Are these adaptations like the originals? Has the writer made changes in dialogue or scenes similar to the ones we find in "Biloxi Blues"? Discuss these changes in class.

9. Look at other playwrights, such as Ibsen, Strindberg, or Pinter, and compare their plays to adaptations based on their works. Can you list other playwrights who have been adapted for the screen?

10. Watch several contemporary movies. Can you find the symbols used to convey the meaning of the piece? Discuss them with members of your class.

11. Read the play and screenplay for "Driving Miss Daisy." Can you find differences in the screen adaptation? Discuss, in class, the reasons for these changes. Are any elements similar? Discuss the reasons the writer chose to retain these elements.

10

Dialogue, Starting to Write, and Marketing

DIALOGUE

A problem many writers have when writing their screenplays is dialogue. Writing good dialogue is difficult, because novel or short story dialogue doesn't easily make the transition into screenplay dialogue. Often, the dialogue is one-sided—characters talk at each other, but not to each other. Dialogue in a novel is often followed by a description of a character's thoughts or other narration, not necessarily additional dialogue.

In the screenplay, the interaction between two characters is essential. Thus, the best approach for writing dialogue when adapting is to follow the same rules that apply to writing dialogue for an original story. The dialogue should be crisp, reveal character, and connect to other characters' lines.

Characteristics of Good Dialogue

Good dialogue should appear to be natural; that is, it needs to sound like human conversation. It shouldn't duplicate conversation exactly or it will be boring. An interesting exercise to try is to tape a few minutes of conversation between two people, or a conversation between yourself and another person. Listen to the conversation; you will see that it sounds natural but has a lot of dead spaces. Your task as a dramatist would be to rewrite this conversation taking out these spaces, adding conflict, and for the most part, shortening the entire sequence. The writer should achieve a balance in dialogue that sounds real, yet is short and keeps the story moving.

Good dialogue should be connected. When a character makes a statement, the next line should respond to what the character has just said. If you can't think of your next line of dialogue, review the speech you have just written and have the next line respond to the content of that speech.

Connecting Dialogue

The simplest way of connecting dialogue is to repeat a single word of dialogue from the previous line. Sometimes you can connect by repeating the entire line with a different emphasis or meaning coming from the way the line is delivered. Avoid simple repetition without adding new words or new meaning to the next line.

Another way of connecting dialogue is to use interruption. One character who is speaking is about to finish a thought, when another character interrupts and completes the line, thereby connecting the two lines. Connected dialogue not only sounds good, it moves the story along.

Play on words is a device for connecting lines of dialogue in which one character takes the previous line or two and plays with some of the words in the speech. The character may make a clever retort or pleasant sounding reply. In Bridges's *The Paper Chase* (1972), for example, Kingsfield asks a student his name. The student replies, "Mr. Bell, as in Liberty Bell." Kingsfield replies, "I will have to dispense with the pleasure of ringing you further, Mr. Bell" (p. 83). Here is a simple play on words with the use of the word ringing. It's amusing and connects the two lines.

Jargon

Good dialogue should use words that reflect the class, occupation, and character of the person speaking. A construction worker uses different words than a doctor. A lawyer probably speaks differently than a sanitation worker. A college professor probably has a different vocabulary than her students.

Remember, also, that dialogue is one of the chief means we have at our disposal for revealing characters. The characters' dialogue should be simple. Use contractions whenever possible. Instead of saying "I will not," say "I won't." It makes the speech move more quickly. Don't be afraid to use incomplete or grammatically incorrect sentences. Use dots (ellipses) for pauses. People rarely speak in a grammatically correct fashion, and they pause often.

If you must write dialects, then tell the actor what dialect you are using and have him or her work on mastering it. If it's a difficult dialect, write one or two lines of your character's speech and give a sample to the actor. Of course, after you've written the speeches, it's always good to have someone read them out loud. Remember, dialogue is written to be heard, unlike dialogue in a novel, which is written to be read. Don't be afraid to rewrite what doesn't sound right.

Keep It Short

A word about length. Students love to write dialogue in which the characters are prone to long speeches. Keep your dialogue short. Make sure dialogue is balanced in a scene. Don't give one character all the good lines and the other characters the one-word responses—keep it balanced. Drama represents both sides of an issue. If, in a scene, both characters have equal points of view, then the scene is more interesting. In a TV interview, Neil Simon stated that he doesn't write villains. When both characters have equal force in a scene, the audience doesn't know who to root for. The audience switches from one character's point of view to the other as the scene develops. It makes the scene and your story more interesting.

Good dialogue serves many purposes. It develops your character, tells your story, develops the issue of the story, and adds conflict. A good *tag line*—a concluding line at the end of a scene—puts a nice finish on the scene. Good dialogue can also be used to connect one scene to the next. "See you at the movie theater." Cut to the movie theater. Good dialogue makes a boring story more interesting and is the heart of your screenplay. Work on hearing the characters speak and don't be afraid to rewrite what they have said. Above all, keep practicing and you'll be on your way to writing good dialogue.

GETTING STARTED

The hardest task for many writers is finding the story to write. If you're writing an original, this task involves a certain amount of research. Remember, every day thousands of unemployed writers are reading the newspaper, listening to the news, and reading magazines looking for dramatic stories. Keep a file box of ideas. An idea that might not seem appropriate at one time can be saved for later use.

For those adapting a novel or short story, the problem of acquiring the rights to the story becomes a factor. Of course, if you use more classic material or material in the public domain, then you can adapt without worrying about obtaining the rights to the story.

However, many of the classics have already been adapted. You might want a more modern story, something that's in keeping with today's commercial marketplace. If you try to adapt a best-seller, you should be prepared to pay a high fee for the right to option the book. An *option* is an acceptable time, a year or two, for which you have the rights, for a certain fee, to adapt another person's work into a screenplay.

Some people say that good novels make poor adaptations and that bad novels make good adaptations. Of course, if a novel hasn't done well commercially, it probably will be optioned for less money. On the other hand, a novel that is a best-seller already has an audience and might make the screenplay easier to sell. The main consideration, apart from the practicality of money, should be the story. Is this a story that would make a good dramatic story with strong commercial possibilities? Are you really interested in adapting this story? Is this the kind of story you want to spend the next six months to a year adapting into a screenplay?

Writing the Screenplay

Once you have decided on your story, it's time to begin the writing process. I suggest you begin with a short premise of three paragraphs to see if you have a beginning, middle, and end for your story. Spend time doing biographies of all the story's characters until you know them as well as you know yourself. It doesn't matter if you never use the majority of biographical details you write in your sketch. The more you know your characters, the better your screenplay will be.

Next, develop a long, well-written treatment outlining the plot of your story. Remember to make your treatment exciting and descriptive because, although it's likely you'll have to write the screenplay to sell your story, some people sell stories using the treatment alone. Don't be afraid of finding that your story changes from the original premise. As you develop the characters and plot details in the treatment, the story should change.

Finally, begin to write a *step-sheet*, or outline, of your story. This outline should detail each scene in each act of the story. You might include the plot details as well as the purpose of each scene in order to remind yourself of what dramatic purpose you're trying to accomplish. Some writers put their outlines on index cards so that they can rearrange the scenes to see if one scene works better in one location versus another. Again, don't be afraid if your story changes. It's supposed to change.

You are now ready to write the first draft. I suggest that you set aside a few hours a day to write the draft, preferably the same time each day, to get yourself into a routine. Remember, writers find new excuses daily not to write. Be disciplined. Follow a routine. When writing the first draft, it's important not to be too critical. It's more important to get material, preferably good material, down on paper. You can always go back and fix things. You'll find it easier to rewrite from a page with dialogue than a blank page. Besides, if you're too critical during the first draft stage, you'll constantly be going back and reworking your material, and you may never finish.

Opinions and Changes

Don't be in a hurry to finish your story. Enjoy the writing experience. Let the writing take you inside the story and the heads of your characters. When you finish your first draft, show it to people whose opinions you respect. Listen to their comments without becoming defensive. Only you know if what they are saying has any validity. If you don't agree, don't change the story. It's your story after all.

If you try to change the story based on every negative comment you receive, you may never finish. Be selective. Use the second draft to rewrite the scenes and dialogue from the first draft that have been bothering you. Don't be afraid to change your story and do the rewrites. Neil Simon's first play, "Come Blow Your Horn" (Simon, 1980), went through ten rewrites before it was finally sold.

Format, Research, and Scriptwriting Software

Writing in the proper format is essential for the beginning scriptwriter. Anyone attempting to sell one's work needs to be sure that it is written in the correct format. Whether one is writing an original screenplay, an adaptation, a sitcom, an episodic, or even a script for corporate or educational TV, correct format is essential.

Whenever possible the beginning writer should attempt to get copies of other scripts dealing with the same subject. If you're writing a spec script for *Seinfeld*, copies of previous *Seinfeld* scripts would be very helpful. When attempting to write movie scripts, the beginning writer should attempt to read as many movie scripts as possible. Scripts can be found in various libraries, purchased at bookstores, and sometimes downloaded from the Internet.

Another useful tool is a group of dedicated software programs that actually put your script in the correct format. If you're writing a sitcom, the software underlines the slug lines, double spaces the dialogue, and capitalizes the action lines. If you're writing a movie script, the software capitalizes the slug line, centers the character's name, and wraps the dialogue.

Depending on the type of computer you are using, there are various software packages available on the market. Final Draft is a popular one for Macintosh users. Moviemaster and Scriptware are two popular ones for IBM-compatible computers. Scriptmaster is good for people interested in writing double-column scripts for radio or corporate scripts. If the cost for these products is too much (averaging around $295), one can modify a word processing program, such as MS Word, by developing a style sheet that can be used for a specific type of script. The Writers Store (Santa Monica Blvd., West Los Angeles) is a great source for all the different types of software on the market.

SELLING YOUR SCREENPLAY

Each year hundreds of budding writers come to Los Angeles to be screenwriters. The odds against making it can be great. If you have perseverance, however, and are prepared to keep working at your craft, the odds can be improved. Certainly you're not going to become famous overnight. Lawrence Kasden, writer of *Raiders of the Lost Ark* and *Body Heat*, took 10 years to sell his first screenplay. Explore alternative markets such as PBS or cable. Not every story that is written has to be a feature for the big screen.

It's important to keep writing. Maybe the first or second screenplay you write won't sell, but it might prove a useful sample of your writing abilities. Many people claim to be writers, but very few have a completed screenplay to show as a sample of their work. Remember, the movie industry is always looking for new material. The key is to keep writing and to lower your expectations that everything you write will sell.

Chances are your first or second screenplay will get you an assignment writing an idea for a producer or adapting a novel a producer has optioned. Ultimately, if you have a produced screenplay that does well at the box office, you will be able to pull out those old screenplays on your shelf and sell them when you become a hot writer. Patrick Shanley, author of *Moonstruck*, subsequently did a movie called *Joe versus the Volcano*. It doesn't seem to be up to his usual high standards. Perhaps it represents one of his earlier screenplays that he recycled after he became a big Hollywood writer.

Agents

For the new writer with a few writing samples, the next step is to find an agent. The Writers Guild West (7000 W. Third St., Los Angeles, CA 90048; 213/951-4000) has a list of all the agents who read material from new writers. The Writers Guild will also register your script, for a fee, to protect your material as you show it around. In addition, you can always use the poor man's copyright by sending a copy of a script to yourself in a sealed envelope. Keep it sealed on your shelf as proof of the date you wrote your script.

Agents are a necessary step in the selling process. They may, if they think you have a hot property, send your material out to different production companies. They may get you invited to meetings with producers who will listen to pitches for your new stories. Having an agent gives you the look of a professional. But, even with an agent, chances are you'll have to do a lot of the contact work yourself. This means getting whomever you know with connections in the industry to read your material. If what you have written is good, someone in a position to do something for you ultimately will read it.

Patience and Success

The key is to be patient. Of course, living in the Los Angeles area is a marked advantage for a scriptwriter because this is where all the work gets done. But for those who don't live in or near Los Angeles, explore your local cable market or PBS station. Many PBS products are produced outside of the Los Angeles area. Don't be afraid to send your material to production companies in Los Angeles even if you don't live there. Some of my students, for example, have submitted their treatments to story departments at various studios and have had their stories optioned. For the most part, however, the story department is there to filter out the mass of material that bombards the studios daily. Don't let this discourage you.

Whenever possible, start with someone higher up in the hierarchy, someone who has the power to make a decision about your work. Of course, if you don't know anyone, it can't hurt to send your material to a studio's story department. Just don't be too disappointed if you don't hear back from them. Rejection is the name of the game in Hollywood and you should gear up to receive your share. If your writing is good or if you can improve your writing skills, chances are you will succeed. Hollywood needs talented people and new and exciting ideas.

SUMMARY

In your quest to become a screenwriter, the key is to start writing and keep writing. Writing is a skill that can only be learned through continued practice. But remember, only you can make it happen. The key is to begin to write. Start now! Have perseverance! And enjoy the adventure that lies ahead.

11

Love Stories

The Bridges of Madison County and *Leaving Las Vegas*

Ron Bass's adaptation of Robert James Waller's *The Bridges of Madison County* (1992) is a prime example of an adaptation which maintains the essence and feel of the original work while making substantial changes in structure, characterization, and social issue to dramatize and visualize this romantic love story of two people who find their "one moment of certainty" through a four-day romantic encounter. This chapter examines this adaptation process by focusing on the following key points.

1. *Find the starting point.* How does the screenwriter find another starting point that differs from the starting point of the novel, and why?
2. *Change in characterization.* How are the roles of the central characters changed in the screenplay to serve the needs of a tighter dramatic structure?
3. *Develop minor characters and a more involved subplot.* How does the writer use subplot to illuminate key issues in the main story line?
4. *Condense scenes and add scenes.* How does the screenwriter change the structure of the novel and for what purpose?

THE BRIDGES OF MADISON COUNTY

Starting Point

Waller's novel begins with a short prologue, a story within a story, a common storytelling device for fiction writers since the days of Moll Flanders. The novel begins with how the story of Robert and Francesca was brought to Waller by the main character's son and daughter. Like other storytellers who went before him, Waller's use of this narrative device sets the stage for the story, hooks the reader into the novel, and allows the writer to comment on the importance of the material he is about to reveal to us. It also sets up the importance of the main character, Robert Kincaid: "He is an elusive figure . . . and was the most challenging part of my research and writing" (Waller, 1992, p. 1).

The first chapter, entitled "Robert Kincaid," begins with Kincaid's journey from Washington State to Iowa. This journey introduces him as the main character in the novel. He wears faded blue jeans, smokes Camels, and wears orange suspenders. "Robert Kincaid was as alone as it's possible to be—an only child, parents both dead, distant relatives who had lost track of him" (Waller, 1992, p. 3). This first chapter not only gives us extensive backstory about Robert but also reveals his philosophy and the real poetic essence and independence of the character.

Robert Bass's first draft of the screenplay, on the other hand, begins with a scene involving Francesca on a train. It is brief but connotes a woman about to begin a new journey. The screenplay then cuts to the farm in Iowa where we meet Carolyn and Michael, Francesca's grown children. The setting and time are sometime after Francesca's death, with the children listening to the stipulations in the will regarding Francesca's wish to be cremated and have her ashes scattered over one of the nearby bridges. This request is startling to the children. Michael chalks it up to some form of dementia and threatens not to comply with his mother's request. Later Carolyn finds a letter that explains Francesca's behavior. This marks the real starting point of the movie.

In the novel, Carolyn's discovery of the letter occurs at the end of the book with the story unfolding in a more typical and linear manner. In the screenplay, the story is basically told as a flashback with Michael and Carolyn reading Francesca's journal of the events. Throughout the movie the story flashes from the past—the love story of Francesca and Robert—to the present, the children's reaction to their mother's story.

I think the first-draft version, although somewhat different from the final shooting script, is significant in giving us some insight into the screenwriter's thinking. By beginning the movie with Francesca rather than Robert as in the novel, the writer is making Francesca the protagonist of the story. In this way, the screenwriter is able to have a tighter dramatic story by focusing on one main character, giving her a problem, Robert, and a conflict (should she go with Robert or stay with Richard, her husband) that will be resolved by the end of the story with the protagonist experiencing character growth (she chooses responsibility over love).

By using the flashback structure as the starting point for the movie, the screenwriter is using several dramatic elements. First, by developing the minor characters of the children, who were barely mentioned in the book, the writer develops a more complex subplot. Both children are involved in unfulfilled marriages, which act to parallel and comment on Francesca's own marriage throughout the course of the movie. Carolyn explains, for example, that she can't believe she's been in a loveless relationship for the last 20 years. But, little did she know that she was taught this by her mother. What other kind of relationship did she ever see while growing up? Interestingly, Carolyn's relationship with her husband results from Francesca's resolution of her conflict. She chooses to remain with her husband Richard out of a sense of loyalty and a decision to do what is morally right. The screenwriter, unlike the novelist, shows us the consequences of this decision, if not on Francesca's own life, then on the lives of her children.

Another aspect of this starting point and repeated flash forwards to the children is the development of character growth as a dramatic element at the end of the screenplay. Normally the protagonist experiences character growth as a result of the resolution of the conflict of the story. In this case, however, in both the novel and the screenplay, the question of how Francesca will resolve her conflict is never really at issue. We assume from what we know of her character that she will stay with the conventional Richard rather than choose the love and danger of the independent Robert. She herself tells us this when confronting Robert over the issue that their love cannot survive if they were to remain together. At least by staying with Richard they will always have their four days together. So the question of growth for her is really a moot point. However, both the children, surrogate representatives of her, make changes in their own marriages following the revelation of Francesca's story. Carolyn chooses to leave her husband and remain on the farm in Iowa at least for a while; and Michael decides to recommit to his relationship with his wife and try to make her happy, realizing that he is still in love with her. Both experience the character growth that the screenwriter could not show us in Francesca.

Change in Characterization and the Development of Conflict

The major change in characterization in *The Bridges of Madison County* is the change in focus of the main character. With Francesca as the main character instead of Robert, the screenwriter can develop the protagonist's problem; create additional conflict; and, most important, give the protagonist something that is at stake if she makes the wrong choice. In this case, Francesca is risking her marriage and her children. Robert as the protagonist has less at stake since he has no family, is used to being alone, and travels around the world leading a fairly interesting life. Screenwriters often change the focus of their original story to achieve a more conventional dramatic structure. In *Leaving Las Vegas*, for example, the screenwriter changes the emphasis of the novel, making Ben the protagonist. In this way, Sera becomes the complication in Ben's life which adds to the problem of his drinking. In the novel, the story begins by placing the emphasis on Sera.

By making Francesca the central character in *Bridges*, the screenwriter has more potential for developing conflict in his story. In the original novel, the relationship between Francesca and Robert is quite calm; obviously, it is love at first sight. Both characters seem to admire each other and respect each other. There is very little conflict. In the screenplay, conflict is essential for holding the audience's interest and moving the story along. Thus, the screenwriter adds conflict between the two characters early in the movie when Francesca challenges Robert's lifestyle.

Francesca is jealous of Robert's previous relationships. "Let's get this all on the table. Now, all over the world you have these conquests, right." Robert responds, "Do you enjoy that thought. . . . All women are easy to damage. . . . Some go looking for it" (Bass, 1993, p. 70). In the novel, Robert admits that he is no monk and Francesca accepts his past and doesn't question him. But in the movie, following her initial sexual encounter with Robert, much more is at stake for Francesca: "These women friends of yours all over the world. Do you write to them? . . . See them again? I just need to know your routine." Robert responds, "There's no routine. I was honest with you." But for Francesca, Robert has an independence that she can never break. She needs to know her affair is different for him. Instead she worries about what other housewives he might be with when he's traveling in other exotic countries on photo shoots.

Later in the movie when Francesca has fallen in love with Robert, the conflict changes. Again, in the novel, there is really very little conflict between the two characters even when Francesca has decided not to go with Robert but to stay with her family. Robert basically accepts her decision and offers to be there for her if she ever needs him. In the movie, there is more tension. The screenwriter adds this entire section dealing with family values and basically gives the two characters something to fight about. "When a woman makes a choice to marry and to have children it's like in a way her love begins and in another it stops. When your children leave they take your life and its details with them.

. . . You never expect love like this to happen." Francesca is afraid; she feels guilty. She tells Robert if they go off together they will lose their love. "Love won't obey our expectations. That wouldn't last if we were together." Robert has the tag line in the scene: "This kind of certainty comes but once in a lifetime" (Bass, 1993, p. 81).

The screenwriter adds additional conflict through the use of the subplot characters of Michael and Carolyn. They continue to argue over their mother's behavior, with Michael acting the part of the dead father, conservative and full of disapproval. Carolyn, very much like her mother, is more understanding. Michael tells her that he can't see his mother having sex. "Mom shouldn't want sex anymore because she's got you." He's jealous of his mother's love for someone other than himself. "If she was so unhappy, why didn't she leave?" **[Exercises 1–5]**

Change in Scenes

As in all movies, scenes must be added or omitted from the original novel, usually to speed the story along. In *The Bridges of Madison County* we have the scene of Richard calling Francesca, which is added in the movie. This scene adds to the pressure on our protagonist showing us how she is already caught between Richard and Robert. Also, in the novel, Robert and Francesca visit several different bridges in the area on their first encounter together. These trips are condensed in the movie—the couple visits only one bridge.

Some scenes are added to the movie for their visual effectiveness. Francesca looking at herself naked in the mirror before writing her note inviting Robert to dinner is added. We realize that she is thinking of herself as a women again for the first time in many years. In another scene we see her being cooled off by a summer breeze. Again, this is a visual moment which suggests her need for the cooling off of her budding passion for Robert. The use of intercuts, also added in the movie, is visually effective. The cuts of Francesca lying in bed while Robert approaches the bridge to find her note have been added. It builds suspense and shows us that her excitement about Robert is building.

The dance with Robert in the kitchen has the added visual element of a shot of Francesca's wedding ring on her finger as she dances. The cut to the subplot involving the son, which shows his intense anger at this moment in the story, has also been added and is effective in showing what the husband's reaction might have been if he had witnessed the same scene. Another addition to the movie involves the characters' declaration of love. In the novel, it happens quickly, almost spontaneously. In the movie, it is more drawn out. It takes longer to happen because the writer needs more time to develop the protagonist's conflict.

The climax scene with Richard and Francesca in the car also has some subtle changes. Visually, we see Francesca with her hand on the door ready to bolt from the car. This moment has been added. Then, in a nice added touch, we see Francesca crying as she realizes she is trapped and cannot leave Richard. At this moment, Richard turns on the farm report on the radio. This added moment says it all visually. Francesca has condemned herself to the boring life of husband, children, and farm reports. The dying scene with Richard has also been added from the novel. It is a fitting resolution to the movie. We see Francesca does love Richard and that perhaps her choice has been correct.

Although *The Bridges of Madison County* is a relatively short novel, scenes are also condensed and cut during its adaptation into the screenplay. The initial scene of Francesca and Robert walking outside after their first dinner together is condensed from its version in the novel. The scene of Robert arriving at Francesca's house, talking to her in the kitchen, and then going to take a shower is also condensed by moving from the arrival of Robert to his taking a shower in Francesca's bedroom. These cuts speed the story along and allow the writer to keep the story moving forward. **[Exercises 6–8]**

LEAVING LAS VEGAS

Like *The Bridges of Madison County*, John O'Brien's *Leaving Las Vegas* is basically a story that centers on two characters. As for *The Bridges of Madison County*, the transition from novel to screenplay involves the classic changes one would expect in the adaptation process: changing the starting point, eliminating backstory, condensing and eliminating scenes, changing the order of scenes, and eliminating characters.

Starting Point

As a love story, *Leaving Las Vegas* has many similarities to *The Bridges of Madison County*. Both stories have negative endings, and both stories change the protagonist. In *The Bridges of Madison County*, the shift of protagonist from Robert to Francesca results in increased conflict and character growth. In *Leaving Las Vegas*, the novel begins with Sera. In essence the novel is her story with the bulk of it devoted to her history, how she got to Las Vegas, her affair with Al, and then her relationship to Ben. Although Ben appears in the novel on page 59, his meeting with Sera on the strip in Las Vegas doesn't even appear in the novel until page 128.

Sera's problem in the novel is her relationship with men, symbolized by her running away to Vegas to escape her relationship with Al.

When she meets men, her dramatic need to have someone who will listen to her is answered with Ben even though his drinking and behavioral problems really intensify her isolation.

In the movie, the story begins with Ben. We learn more of his backstory, such as the loss of his job, his break-up with his wife, and so on, through a series of flashbacks interwoven into the story. Backstory on Sera is kept to a minimum. The problem in the story then becomes his desire to escape his past through drinking and ultimately death. The relationship with Sera serves only to complicate this problem. Although the movie's emphasis on both characters is important, this shift to Ben serves several purposes. Of the two characters he seems the most redeemable—a drunk over a prostitute—with his backstory providing more acceptable motivation for an audience's sympathy than Sera's. In addition, alcoholism as a social issue has more potential for this story than prostitution. **[Exercises 9–11]**

Elimination of Backstory

With the change from Sera to Ben as the main character of the screenplay, the writer is able to eliminate the large amount of backstory that the novelist (O'Brien, 1990) provides on Sera: "Once a small girl in the east, she now lives here. There was a time spent in Los Angeles . . . and she wishes to stay here in Las Vegas, where she arrived long enough ago that she now calls it home when speaking with herself" (p. 2).

The character's ghost, her relationship with Al, is seen in the movie but the reason she is running away from him, which is explained in the novel, is eliminated—"She was haunted, pursued, tortured emotionally, sometimes physically, day and night by the one who had made her the object of his obsession. She was and would become his last, best gold chain, an unwilling bauble on his furry chest" (O'Brien, 1990, p. 19). Al's violence and the incident when he kicked her in the stomach while with another women is also omitted. Like her history with Al, her relationship with her father and the incident with an ice cream cone (on page 28 of the novel) is omitted from the movie.

Even with the main protagonist in the story, it is necessary to condense the large amount of character history the novelist is apt to give us. Thus, Ben's previous relationships with prostitutes are omitted. The incident with him trying to pick up a woman in a bar and inviting her to have sex with him because he is such a great lover is also cut. The incident on the 405 freeway, when he is stopped by the police at 4 A.M. after a night of heavy drinking also is omitted.

Interestingly the introduction of Ben into the story is also changed from the novel. In the book, we meet Ben in a bar at about 10 A.M. drinking, of course, and watching a quiz show on TV. In the movie, the writer goes for a more visual opening of Ben in a supermarket loading up an entire shopping cart full of liquor. This is more effective as an opening because it really shows that the character has some major type of problem. In the book, Ben's drinking in a bar could just be a normal occurrence. No one, however, buys a bottle of every type of alcoholic beverage possible at one time. **[Exercises 12–13]**

Elimination of Characters

With the emphasis in the movie switched to Ben, several of the characters that surround Sera are eliminated. Sabrina, for example, the 16-year-old prostitute who lives with Sera for a short time, is cut out of the movie. In addition, the characters of Mary and Slim, Sera's neighbors, are cut. "She was doing her best at maintaining a light friendship and though she was indeed being deceptive about much of her life—a deception which caused her some distress—she had some affection for Mary and Slim" (p. 37). Thus, we lose the details of Sera's friendship with Mary, Slim's flirtations with Sera, and the final breakup of the relationship when Mary learns that Sera is a prostitute. Again this detailed development of the subplot characters that surround Sera is unnecessary with the focus of the story centered on Ben.

In addition to eliminating the characters surrounding Sera and backstory about Sera, the writer cuts down on the detailed explanation about her actions in the movie. Again the screenwriter opts to give us information visually rather than giving us a written explanation, as in the novel, for a character's action (O'Brien, 1990, p. 6):

> And she is a good thing, good at this thing. Paying for and using her, there are always men available. The tricks turn to her for she glistens with the appealing inaccessibility of the always introspective. They turn to her for the buyable quench—no lie, a promise in the panties—and she plays out the bargain with the competence of one consistently able to hit well the mark. No matter how long it has been since she last worked, she is never without a full arsenal of idiosyncratic performance ready in an instant to fall into the groove and take command without missing a beat.

The screenwriter shows us Sera's ability in scenes with men who buy her services. The novelist talks about it, explains it, gives us the real insight through words that the screenwriter must do through pictures.

Omission of Scenes

The prime directive for the screenwriter is always to keep the story moving along. To accomplish this task, scenes that may be interesting and shed additional light on the motivation of the characters must often be eliminated. In *Leaving Las Vegas,* several scenes from the novel have been cut. In the opening section dealing with Sera, we lose the scenes of Sera gambling, her trip to the Fremont Hotel, the incident with the man at the bar, and her ultimate expulsion from the bar. The scene with her in the cab driving down the streets of Vegas while the driver listens to the call-in show on the radio has also been cut.

In the section about Ben, several key scenes in the novel have been omitted. The entire routine Ben goes through each day, starting at certain bars that open at certain times, his preference for bars in Beverly Hills, and his constant problem of trying to maintain his alcoholic high are omitted. "At six the hard-core bars opened and the stores can sell, though they sometimes choose to withhold . . . nine A.M. is considered a safe opening time for the bars that don't like to admit that people drink that early" (p. 64). Ben's routine in Los Angeles is interesting but the focus of the movie is his time in Vegas with Sera and the screenwriter moves to get him to Vegas as soon as possible, eliminating Ben's notice to his landlord, his packing and the demise of his various possessions.

Ben's motivation for moving to Vegas to have access to 24 hours of open bars is barely touched on in the movie. "The inconvenience of having to keep track of time is steadily growing as his inclination and even his ability to do so diminishes" (p. 88). Ben's motivation and drinking habits are unnecessary for moving the story along. The writer focuses instead on Ben's meeting with Sera and the development of his relationship with her, which will add to his conflict and problem. **[Exercises 14–15]**

SUMMARY

Love stories have always been a strong genre for the Hollywood dream machine. More than other types of stories, conflict and character growth are two key ingredients to the successful dramatization of love stories. In both *The Bridges of Madison County* and *Leaving Las Vegas*, the structure and emphasis of the novels have been changed to provide this conflict. Both screenwriters have not been afraid to modify the books by changing the importance of the novels' characters. In *Leaving Las Vegas* this change is more effective than it is in *The Bridges of Madison County.* Unfortunately, sometimes the changes a screenwriter makes don't really improve the story.

EXERCISES

1. Create a male and female character. Describe them. Have them meet and describe their first encounter, making it suggestive of future romantic involvement.
2. Make the characters gay and rewrite the scene using two male and then two female characters.
3. Write a climax and break-up scene for the characters in exercises 1 and 2.
4. Describe two love stories, the characters' problems, and how they were resolved.
5. Repeat exercise 4 for two romantic comedies.
6. Rewrite the ending of this movie as if Francesca had gone away with Robert. How would things have been different?
7. Discuss the endings of several love story movies. Compare ones that ended positively to ones that ended negatively. Which ones were more effective and why?
8. Can you name and discuss the endings of any love stories that ended ambiguously? For example, look at the ending for *She's the One*. What type of ending did the writer use? Was it effective?
9. Discuss another novel in which the protagonist is changed from the novel to the screenplay.
10. Examine another movie that deals with alcoholism as a social issue. How is the protagonist made sympathetic?
11. Discuss another movie that deals with prostitution as an issue. How is the protagonist made sympathetic?
12. Discuss two movies that begin with visual openings to illustrate the protagonist's problem.
13. Discuss two movies that open with action-oriented beginnings.
14. Write a scene in which Sera tries to talk Ben out of drinking by using her feelings toward him as a way of persuading him.
15. Take these two characters and make the problem surrounding them involve another addiction such as drugs or sex. Rewrite the scene.

Appendix

What follows are suggestions and answers for some of the writing exercises in the chapters. Instructors may want to assign some other exercises to those students who intend to go beyond the normal range of the course. Some instructors also may want to supplement these answers with handouts of specific scenes or story excerpts.

CHAPTER 2: PRINCIPLES OF DRAMATIC WRITING

1. Finding act breaks can be difficult unless one is viewing the movie for the second and third time. Usually the first act break is easy to find. By the time we reach the second act break, we are so deeply into watching the movie that we forget to look for the act break. Watch a movie for the first time for enjoyment. Watch it a second or third time for analysis. Good movies can be watched over and over and still yield fresh insights. In a two-hour movie, the first act break comes 20 to 25 minutes into it. The second act break comes around 90 minutes into the film.

In *The Paper Chase*, the first act breaks when Hart oversleeps and is thrown out of the upper echelon about 25 minutes into the movie. The second act break occurs when Hart fails to deliver the research Kingsfield asks him for around 90 minutes into the movie. In *Harold and Maude,* the first act break occurs when Harold's mother realizes he has been scaring his first computer date. The second act break occurs when she threatens to have Harold inducted into the Army.

In *Kramer versus Kramer*, the first act break comes when Ted gets the letter from Joanna officially announcing she won't be coming back. The second act break occurs when she returns to New York and tells Ted she wants Billy back. After both breaks, the plot moves in a new direction. After Act One, we see Ted throwing out Joanna's possessions and starting his life over. After the Act Two break, we see Ted consulting a lawyer and planning his case.

2. In half-hour and hour shows, the act breaks are more evenly spaced than in the two-hour movie, in which the second act is by far the longest.

3. Twists or unusual endings are nice for stories, but certainly are not mandatory. Depending on the genre of the story you are writing—mystery, comedy, or a love story, for example—twists make for more interesting endings. If your story lacks a twist, at least make the protagonist's solution believable. In the movie *Angel Heart*, the twist at the end occurs when the protagonist, a detective, discovers that the man he's been searching for is himself and that his client is the devil. He made a pact with the devil that he has blocked out of his mind. When he realizes the truth, the devil comes and takes him away. In *Kramer versus Kramer*, the twist comes at the end when Joanna decides to give Ted custody of Billy even though she has won the case.

4. Choosing a character's name can be a device for describing your character. Names should be interesting and fit the character you are writing about. In this exercise, I have deliberately picked stereotypes to illustrate how a name conjures up an immediate mental image of the character. Some writers use stereotypes, but go against the traditional associations that go with a particular name. This can make for interesting characters and even create internal conflict between how people view a character and how he views himself.

5. Watching as many contemporary and classic movies is quite helpful for seeing how other writers create and reveal their characters. Usually it requires more than one viewing to pick up on the details required in this exercise. In *Kramer versus Kramer*, Ted is the protagonist. We see his main character trait, being a workaholic, in the scene that shows him staying late at work. His goal in the story is to raise Billy and retain custody. Joanna might be considered the antagonist because she threatens Ted's goal. But the writer makes her sympathetic as well. We see how much she cares for Billy in the opening scene when she is talking to him while she packs her things.

6. The background of a character can be a very useful device for creating conflict. People who have been brought up with different values are usually going to have different goals in a story. This difference will lead to conflict. In *Driving Miss Daisy*, the very different backgrounds between Daisy's Jewish upbringing and worldly travels conflicted nicely with her

chauffeur's. Not only was he a southern black, but he also had a rather sheltered upbringing—during his 65 years, he never left the state of Georgia.

7. Usually, characters' long speeches come in the third act of a story when the character breaks down and reveals in a long speech, but not a theatrical soliloquy, her real passion for the problem she faces in the story. In *And Justice for All*, the actor playing the lawyer protagonist (Al Pacino) gives a long speech in which he condemns Judge Fleming, the movie's antagonist, during the climax of the movie. He not only reveals Fleming's guilt, but also the guilt of the entire legal system.

In *Kramer versus Kramer* in the courtroom climax scene, Ted Kramer gives a long speech in which he talks elegantly on the rights of fathers to have custody of their children. He asks what makes a mother a better parent by virtue of her sex. Long speeches can be quite moving and helpful for revealing the writer's theme in a movie.

8. Premises are useful devices for organizing a potential story idea. Of course, the whole process of pitching premises at a story meeting is an oral one. The premise is useful for getting a story editor or producer interested in and excited about your idea. The premise should never have too many story details. That's the purpose of your treatment. The premise sells your idea. The treatment is a written version that outlines all your plot details and ideas for characterization. In a step-deal in which the producer can cut the writer off at any stage, a well-written treatment is essential for getting to the next stage of writing a first and second draft.

In the premise, the first couple of sentences should introduce your protagonist, his or her initial situation in your story, and should lead up to the character's problem at the end of Act One. The second paragraph should include how the protagonist solves this problem, additional complications, and the crisis at the end of Act Two. The third paragraph should include the climax, twist, and resolution. Each paragraph representing each act should only be a few sentences long.

9. Preparation is essential to the well-written story in order for the audience to believe the protagonist's solution to her problem. Foreshadowing is the writer's tool for subtly planting this preparation so the audience believes the ending to the story. In *Greystoke*, the story of Tarzan and his return from the jungle to England, the writer introduces the character of Greystoke's grandfather. In an early scene, the grandfather warns Tarzan that the steps in the family mansion are slippery. Later, when the grandfather falls and dies on these steps, we are not surprised.

CHAPTER 3: CHARACTERIZATION

1. This rewrite requires dropping the sarcasm from Ben's speech and making the father less understanding of Ben's situation. In the

screenplay, the father doesn't hear Ben when Ben tells him he doesn't want to meet the guests at the party or go to graduate school. The father is wrapped up in his own little fantasy world about Ben's future. This change immediately makes Ben a more sympathetic figure in Henry's screenplay (1967, pp. 5, 7) than he was in the novel:

```
                         MR. BRADDOCK
          The guests are all downstairs, Ben. They're
          all waiting to see you.

                         BEN
          Look, Dad—Could you explain to them that I
          have to be alone for a while?

                         MR. BRADDOCK
          These are all our good friends, Ben. . . .

                         MRS. BRADDOCK
          They came all the way from Tarzana. Now
          let's get cracking.
```

2. When Charles leaves his office and passes by the secretarial pool, he picks up a vase of flowers and puts them on one secretary's desk, he plays with the transcribing earpiece of another secretary, and he sticks a pencil through a third woman's large afro. These actions usually evoke a laugh from most audiences watching the film; they make the character likable. He has a sense of humor. Note there are three actions. Why? Usually, there are three acts to a story. Events seem to occur in threes. It takes three to build up and pay off the final laugh in this scene.

3. In *A Clockwork Orange*, society, not Alex, is the antagonist of the movie. Even though we see Alex raping and killing in the movie, the character has an interesting personality. We can understand why he's so violent, because everyone in the movie, society as a whole, is violent. Despite Alex's violence, the writer keeps him likable due to his interest in classical music, his sense of humor, and his interesting use of the English language.

4. Sargent's (1979, pp. 109–10) scene with Jeanine and Conrad reveals Jeanine's belief in God:

```
                         JEANINE
          Did it hurt?

He realizes what she's talking about. He's not ready for
that.
```

```
                    CONRAD
No. I don't think so. I don't remem-
ber . . .

                    JEANINE
Do you believe people are punished for the
things they do?

                    CONRAD
I don't believe in God.

                    JEANINE
Not at all?

                    CONRAD
It isn't a question of degree. Either you
do or you don't.

                    JEANINE
I believe in God.

                    CONRAD
So what are you afraid he'll punish you for
something?

                    JEANINE
I've done a lot of things that I'm ashamed
of.
```

Unlike the character in the novel, Jeanine barely reveals her sordid past; it's not necessary. The story isn't about her, it's about Conrad.

5. As Paddy Chayefsky said, everyone's life has the potential for drama. Even the incidents in your background might make for an interesting dramatic story.

6. When revealing backstory, it's important to do so subtly. Don't come right out and state, "This is backstory about my character." Work it into the conversation so that it's not obvious that you are revealing backstory. In *One Flew Over the Cuckoo's Nest*, the writer includes a scene between the protagonist (Jack Nicholson) and a psychiatrist. The doctor asks how the protagonist happens to be there. The protagonist reveals his backstory concerning his affair with a 16-year-old girl.

7. Here are Sargent's (1979) descriptions of Calvin and Jeanine:

```
"CALVIN JARRETT, 40, . . . a decent-looking, middle-class
businessman. Intelligent. Friendly." (p. 3)

"JEANINE, a very pretty girl." (p. 1)
```

8. Be selective when picking objects in the rooms of your characters. They may be your only device for revealing character. In *The First Deadly Sin*, the killer's apartment contained objects that made it appear sterile and clean, two key character traits for this particular character.

9. Good description appeals to all of the senses. In a screenplay, the writer usually focuses on the visual and sound description to convey the image.

10. Here is a sample from a student script dealing with this exercise. This was written by Denise Rice in RTVF 225 on May 9, 1990:

```
INT. CLASSROOM—NIGHT

Mrs. Marmalard, a middle-aged woman wearing out-of-date
clothes, more make-up than Tammy Faye Baker, and smelling of
cheap perfume, slowly approaches the door with her daughter
Laurie, who, at 16, looks like she's still going through
puberty.

                           LAURIE
              Mr. Stone, this is my mother, Mrs. Beatrice
              Marmalard.

Mr. Stone, an over-the-hill reject from the sixties, strolls
very calmly over to the two women.

                           MR. STONE
              Hello, please let me show you around. . . .
              This is one of my favorites . . . two
              people embraced in a hug. The statue shows
              such sincerity . . . forgiveness and
              understanding.

                           MRS. M.
              It looks to me like a man trying to kiss up to
              his woman for doing something stupid.

She walks over to another statue entitled "The Burning Bed."

                           MRS. M.
              Now this is the one I like. Revenge. The
              perfect motive. Yes, this student is
              definitely going far.

Mr. Stone sighs and stops smiling.

                           MR. STONE
              I can see we obviously have different views
              on our opinions of art.
```

```
                    MRS. M.
That's not the only thing we don't have in
common.
```

11. Actually, in Woody Allen's scene, the writer uses the dialogue between Allen and Bogart to create the comic moments. Allen is afraid to seduce Linda. Bogart is coaching him by giving him the right lines. Allen doesn't believe the lines will work, but they do (Anobile, 1977, pp. 140–41):

```
                    BOGART
Now—tell 'er that you've met a lot of dames,
but she is really something special.

                    ALLEN
Oh, that she won't believe.

                    BOGART
Oh no?

                    ALLEN
I have met a lot of dames—but you are
really—something—special.

                    LINDA
Reeeally?

                    ALLEN
She bought it!

                    BOGART
Okay. Now put your right hand around her
shoulders and pull her close.

                    ALLEN
That I can't do. No, I really . . .
```

CHAPTER 4: STRUCTURE

1. Since this is only a half-hour story, we don't have time to waste. Some people would like to start with a scene of the young boy graduating from high school. Others would choose an opening in which the young man, walking down the street, sees a "Help Wanted" sign and goes in for a job interview. But, to speed the story along, Chayefsky began with the young man already at work in the print shop and the

information about his graduation is treated as backstory and revealed through dialogue in the opening scene.

2. Chayefsky picks a religious man to be his protagonist to intensify the loss of faith. The man, a cantor in a local Brooklyn synagogue, is lying on his bed in the opening scene, bemoaning his loss of faith in God. For further discussion of the choices Chayefsky made in this story, read Chapter 8 on the problems of adapting from the short story.

3. The opening scene of a story should reveal the major conflict and problem of your protagonist. Writers approach this situation in different ways—from the subtlety of the symbol of the cowboy for Joe Buck, in *Midnight Cowboy*, to the more obvious macho revelations in the James Bond movies. In *Kramer versus Kramer*, the opening scene reveals Joanna leaving Ted. This is Ted's problem. The second scene reveals Ted staying late at work. The scenes reveal the conflict that will develop between Ted as a career man and Ted as a father.

4. In the screenplay, the internal dialogue—"What is she talking about?"—is omitted. The parts about the roommate being at the library, the excerpts about looking for work and coming back to work in the library, all the dialogue regarding Sam and getting any job, the parts about him loving her and wanting to know what's going on, how they ate together and went to the movies together are omitted. Basically, in the screenplay, the scene opens with the dialogue about the dessert, progresses to Charles asking Laura to stay with him, and moves to the dialogue about the taxi driver. The screenplay's scene is shorter and moves more quickly.

5. In the novel *Ordinary People*, the ending has Conrad going back and again becoming friends with Lazenby. In the screenplay, it ends with a scene that reestablishes the relationship between Conrad and his father. The screenplay focuses on the main relationship in the book—the one between Conrad and his father. It can't wrap up all the minor ones as well. In the novel *Out of Africa*, the story ends with the protagonist talking of her friends among the natives and what has happened to them. In the screenplay, the ending focuses on the character of Denys Finch Hatton and how the protagonist has heard that lions are seen to come frequently to his grave. Once again, the screenwriter has chosen to focus on an ending that reveals information regarding the most important relationship in the story.

6. Again, the writer wants to conclude his story by showing us what will happen to the main characters. Melba, a minor character, is omitted from the ending of the screenplay. In the novel, we are reminded of the conflict that existed between Aurora and Emma. In the screenplay, the writer goes for a more positive ending with a shot of Aurora and Emma's son. There is hope for the future. Tommy will come to grips with his mother's death and Aurora will have character growth and raise her granddaughter in a better fashion than she did Emma.

CHAPTER 5: *KRAMER VERSUS KRAMER*

1. Here is Benton's description of Joanna (1978, p. 1):

```
"JOANNA—In her mid-thirties. She is beautifully dressed in a
style that can best be described as Bloomingdales."
```

2. Joanna tells Ted she wants her son; she's had enough of watching him from a distance. Ted reminds her that she's the one who walked out on Billy. Joanna's point is that they had a bad marriage and when she left she was messed up. She only knew how to be a daughter or a wife; she needed to discover who she really was. Ted doesn't care. She can be whatever she wants, even a mother, but she can't have his baby. When Ted refuses to discuss it rationally, he tells her to go to hell.

3. A slightly different emphasis in the story might require a different choice of backstory. For example, if the singles' scene was made more of an issue in the story, you might choose more backstory dealing with Ted's days on Fire Island and how he met Joanna.

4. An example of connected dialogue occurs in the restaurant scene with Ted and Joanna quarreling over Billy (Benton, 1978, p. 89):

```
                    TED
        Come on, Joanna, what did you learn? I'd
        really like to know.

                    JOANNA
        I've learned that I want my son.
```

Note that one device writers use for connecting dialogue is to repeat a single word from the previous line. This is even more effective when the writer changes the word around and connects the new line with some clever play on one of the words from the previous line of dialogue. Sometimes using the same word in three consecutive lines connects three lines of dialogue. To avoid repetition, the writer should add something new to the next line in addition to the repeated word.

An example of foreshadowing might be the scene in which we see Joanna sitting in the window in the coffee shop. We know that she's back in town and that something will happen regarding Billy.

CHAPTER 6: SOCIAL ISSUE

1. Basically, in the climatic courtroom scene, Ted states that he's sorry his marriage broke up, but that's not the issue. He believes that

Joanna loves Billy, but that's not the issue. The issue is who is the better parent for Billy—Joanna or Ted. Ted's position is that being a good parent has nothing to do with your sex, but depends on other qualities, such as patience, communicating with the child, and being there for him. And, of course, he does love Billy very much.

Joanna's position is similar to Ted's. She knows leaving her son was a terrible thing to do, but why shouldn't a woman have the same rights to a life and a career as a man? Joanna contends that she now has her life together and her son needs her. He needs his father, yes, but he needs a mother. The issue of motherhood versus fatherhood is the central question in this story.

Billy doesn't testify because of his age, but it might be interesting to hear his position if it had been written. We can see how he feels at the end of the movie when he wants to see his mother, but feels insecure about leaving his dad, his toys, and the security of his room.

2. Religious differences might make an interesting social issue for exploration in a screenplay. It's important to give all sides of the issue, using different characters to represent the positions.

3. How would your marriage breakup differ from the one in *Kramer versus Kramer*? Would the sex of the child make any difference? Is a boy better off with his father or a girl with her mother? Maybe sex has nothing to do with the issue.

4. Are racial issues legitimate grounds for marital difficulties? Would race play any part in a decision affecting custody of a child?

5. Kevin replies that Hart knows better than that. They may be marks on a piece of paper, but they determine futures and salaries. Brooks concludes that with his grades, it's going to be hard just keeping his family in baby food.

6. Hart comes to realize that grades aren't that important. His rejoinder might be keyed on the words "future" and "salaries," arguing that grades really don't determine these two elements.

7. Your issue might focus on the relationship between good grades and job opportunities or the relationship between a diploma and a future career.

8. These issues are in the news daily. Pick some new aspect of the issue to explore with your characters. Use different characters to represent different sides of your issue.

9. Maude talks about her past as a protester and her love relationships. She talks about an affair she had with a man in Europe. Ironically she was in a concentration camp, yet she still has a positive attitude about life. Harold, on the other hand, is a product of an uncaring mother. He talks about an experiment in chemistry class when the lab exploded. Harold's mother was told he was dead. Her reaction was fake. She fainted, but Harold knew she really didn't care. Now he stages

fake suicides to get her attention. He tells Maude, "I haven't lived. I've died a few times" (Higgins, 1971, p. 61).

10. Watching movies just to discern the social issue can be a good exercise in seeing how the writer uses different characters to frame the positions of the issue. The writer's own view on the issue is usually determined by the way he has the story end. Subplot parallels the message of the main plot.

11. The basic argument is the one we've been hearing throughout the movie. One lawyer contends that abortion is killing a human being and that the fetus is alive from the moment of conception. The other lawyer is arguing for the rights of women to control their own bodies.

12. Since the *Roe vs. Wade* decision, the abortion issue has come up before the Supreme Court. Your argument might reiterate the *Roe vs. Wade* argument and then break new ground by involving the state and its support of abortion through public funds. Check the paper to see how the issue was narrowed down by the Supreme Court in its latest decision.

13. Check the newspaper for current issues. Write down the different sides of an issue and try to create characters who would hold these positions. If you are stuck, you can always use the characters in the newspaper article, but change their names.

CHAPTER 8: THE SHORT STORY

1. I've wanted to see someone try an adaptation of J. D. Salinger's *Nine Stories* or Woody Allen's collection of short stories in *Side Effects*.

2. A more current issue might include a story dealing with space safety or the military use of space.

3. You might compare some of the adaptations of Edgar Allan Poe's stories, such as "The Fall of the House of Usher," or the adaptation of Shirley Jackson's "The Lottery."

4. The adaptation is similar to the short story. The film seems to emphasize the size and power of the rocking horse, and the sounds of the house calling out for money.

CHAPTER 9: PLAYS

1. The screenwriter omits the exchange about Jennie being an actress, the part about the brochure, the occupational hazard, the part about scrutiny and prying, the dialogue about being related and talking in the same rhythm, and cuts to the part about being practical sooner, which is really the heart of this scene. The scene is about bringing these two

characters together for a date. The practical thing to do, George argues, is to get together so that they can get back to their own lives again.

2. Simon, like most adapters of plays, cuts scenes down to make them move, getting quickly to the relevant point while pushing the story along.

3. Plays need to be opened up for the screenplay, such as scenes taken outside and additional scenes included for variety.

4. All the dialogue leading up to the crucial question of getting involved has been cut.

5. Lines about it not being 1934 and getting today's paper are cut. Lines about being a hot property and doing musicals are cut. Lines about living in a five-room suite are cut and lines about wearing the same goddamn pajamas are cut. Lines about taking crap, about saying anything without permission, about owning 50 percent of the act, about going to court with a cow have all been cut. The entire scene is trimmed.

6. Special effects are used in the famous dagger scene in *Macbeth*.

7. Similar condensing of scenes occurs in these plays, along with changes in endings and characterization.

8. Some adaptations are more faithful than others.

9. Symbols should be subtle and never intrude on the screenplay.

10. As with most screenplays adapted from plays, the screenplay has been opened up, scenes shortened, and dialogue cut.

Bibliography

Agel, Jerome (1970). *The Making of Kubrick's* 2001. New York: Signet.
Angus, Douglas (1985). *The Best Short Stories of the Modern Age.* New York: Ballantine.
Anobile, Richard J. (1977). *Woody Allen's* Play It Again Sam. New York: Grosset and Dunlap.
Bass, Ron (1993). *The Bridges of Madison County.* Los Angeles: Warner Brothers.
Beattie, Anne (1976). *Chilly Scenes of Winter.* New York: Warner.
Beja, Morris (1979). *Film and Literature.* New York: Congman.
Belmer, Rudy (1986). *Inside Warner Brothers.* New York: Ungar.
Benton, Robert (1978). *Kramer versus Kramer.* Burbank: Columbia Pictures.
Bluestone, George (1957). *Novels into Film.* Baltimore: Johns Hopkins Press.
Brady, Ben (1974). *The Keys to Writing for Television and Film.* Dubuque, IA: Kendall Hunt.
Bridges, James (1972). *The Paper Chase.* Los Angeles: Thompson-Paul Prod.
Brooks, James L. (1982). *Terms of Endearment.* Los Angeles: Paramount Pictures.
Carter, Wayne (1987). "The Dice Man" (unproduced screenplay). Los Angeles: Paramount Pictures.
Chandler, Raymond (1975). *Omnibus.* New York: Knopf.
Chayefsky, Paddy (1954). *Television Plays.* New York: Simon and Schuster.
Corman, Avery (1977). *Kramer versus Kramer.* New York: Signet.
Craven, Thomas (1926). *The Great American Art.* New York: Dial.
Crichton, Michael (1990). *Jurassic Park.* New York: Ballantine Books.
Cross, Alison (1989). *Roe versus Wade.* Los Angeles: CBS.
Guest, Judith (1976). *Ordinary People.* New York: Ballantine.
Grisham, John (1991). *The Firm.* New York: Island Books.
Harris, Thomas (1988). *Silence of the Lambs.* New York: St. Martin.
Henry, Buck (1967). *The Graduate.* Los Angeles: Lawrence Thurman.
Herlihy, James Leo (1969). *Midnight Cowboy.* New York: Avon.

Higgins, Collins (1971). *Harold and Maude*. Los Angeles: Paramount Pictures.
Kittredge, William, and Steven Krauzer (1979). *Stories into Film*. New York: Harper.
King, Stephen (1982). *Different Seasons*. New York: Signet.
Lovell, Jim, and Jeffrey Kluger (1995). *Apollo 13*. New York: Pocket Books.
McMurtry, Larry (1975). *Terms of Endearment*. New York: Signet.
McPherson, Scott (1992). *Marvin's Room*. New York: Penquin Books.
O'Brien, John (1990). *Leaving Las Vegas*. New York: Grove Press.
Ondaatje, Michael (1993). *The English Patient*. New York: Vintage.
Peters, Charles (1980). "Paternity" (unpublished). Los Angeles.
Rhinehart, Luke (1971). *The Dice Man*. New York: William Morrow.
Salt, Waldo (1969). *Midnight Cowboy*. Los Angeles: Hellman-Schlesinger Production.
Sargent, Alvin (1979). *Ordinary People*. Burbank: Wildwood Enterprises.
Simon, Neil (1980). *Collected Plays, Volume II*. New York: Avon.
Simon, Neil (1988). "Biloxi Blues." New York: Signet.
Sinyard, Neil (1986). *Filming Literature*. Sydney, Australia: Croom Helm.
Stanley, Patrick (1988). *Moonstruck*. Los Angeles: Paramount Pictures.
Swain, Dwight V. (1984). *Film Scriptwriting: A Practical Manual*. Boston: Focal Press.
Turow, Scott (1970). *One L*. New York: Warner Books.
Waller, Robert James (1992). *The Bridges of Madison County*. New York: Warner Books.
Webb, Charles (1963). *The Graduate*. New York: New American Library.

Index